H. A. BUCHAN

IEE SUPPLEMENTARY PUBLICATIONS TO BS 7671 REQUIREMENTS FOR ELECTRICAL INSTALLATIONS

IEE WIRING REGULATIONS, 16TH EDITION

IEE have published a series of Guidance Notes which enlarge and amplify the requirements specified in the British Standard for Electrical Installations. These texts provide an extensive supplement which designers and installers will find a valuable aid to the effective use of the basic document.

GUIDANCE NOTES

Six Guidance Notes are available on

1. Inspection and Testing
2. Isolation and Switching
3. Selection and Erection of Equipment
4. Protection against Fire
5. Protection against Electric Shock
6. Protection against Overcurrent

ON-SITE GUIDE

A convenient short form guide for competent electricians which covers domestic installations and smaller industrial and commercial installations up to 100 A per phase. It removes the need for detailed calculations and underlines the importance of detailed inspection and testing of installations. It provides comprehensive checklists and procedures to guide the competent electrician when dealing with these smaller installations.

BUSINESS REPLY SERVICE
License No. SG 250/20

MARKETING (WRGS)
The Institution of Electrical Engineers
Michael Faraday House
Six Hills Way
Stevenage
Herts
SG1 2BR

IEE SUPPORT FOR BS 7671
REQUIREMENTS FOR ELECTRICAL INSTALLATIONS
IEE WIRING REGULATIONS, 16TH EDITION

TRAINING AND SHORT COURSES

The Institution offers a full range of training for the 16th edition of the Regulations.

TWO-DAY courses to enable those without previous experience of the Wiring Regulations to understand and apply the 16th edition.

16th EDITION DESIGN is aimed at those with responsibility for design or specification of electrical installations. The course is also suitable for those in allied professions who require a broad understanding of the Regulations.

INSTALLATION AND MAINTENANCE will enable those involved in new installations, large or small, or alterations or additions, and maintenance to understand the detailed requirements and practical application of the 16th edition.

DISTANCE LEARNING

There are three distance learning courses available on the 16th edition of the IEE Wiring Regulations in either PC computer format or videotapes with supporting notes and workbooks.

- R01 Design of Installations
- R02 Installation and Maintenance
- R03 An Introductory Course

Please send me further information on the items I have ticked below:

☐ Publications—Guidance Notes   ☐ Short courses—16th Edition Design
☐   —On-Site Guide   ☐   —Installation & Maintenance

☐ Distance Learning—Course R01
☐   R02
☐   R03

Please tick box to receive future issues of *Wiring Matters* ☐

Name ....

Address ....

....

BRITISH STANDARD

BS 7671 : 1992

Requirements for Electrical Installations

IEE Wiring Regulations
Sixteenth Edition

Copies may be obtained from The Institution at the following address:

The Institution of Electrical Engineers,
P.O. Box 96,
Stevenage, Herts. SG1 2SD

© BSI AND THE INSTITUTION OF ELECTRICAL ENGINEERS. 1992
NO COPYING IN ANY FORM WITHOUT WRITTEN PERMISSION

Published by: The Institution of Electrical Engineers, London

© 1992 Institution of Electrical Engineers and British Standards Institution

All rights reserved. No part of this publication may be reproduced, stored in a retrieval system or transmitted in any form or by any means — electronic, mechanical, photocopying, recording or otherwise — without the prior written permission of the publisher.

British Library Cataloguing in Publication Data

A CIP catalogue record for this book
is available from the British Library

ISBN 0 85296 557 5

Printed in the United Kingdom by A. McLay & Co. Ltd., Cardiff

Foreword

This British Standard is published under the direction of the Power Electrical Engineering Standards Policy Committee of the British Standards Institution (BSI) and the Council of the Institution of Electrical Engineers (IEE).

This Standard replaces the 16th Edition of the IEE Wiring Regulations and is identical to it. Copyright is held jointly between BSI and IEE. Technical authority for this work is vested in a Joint Committee which combines the IEE Wiring Regulations Committee and the BSI Technical Committee PEL/64. This Joint Committee, whose constitution meets the requirements of both its parent bodies, takes responsibility for the contents of this British Standard under the joint authority of the Council of the IEE and the BSI Standards Board.

All references in this text to the Wiring Regulations or the Regulation(s), where not otherwise specifically identified, shall be taken to refer to the British Standard, Requirements for Electrical Installations BS 7671:1992, formerly the IEE Wiring Regulations.

Contents

Editions	vi
Wiring Regulations Committee, Constitution as at March 1991	vii
Preface	viii
Notes on the plan of the 16th Edition	ix

PART 1 — SCOPE, OBJECT AND FUNDAMENTAL REQUIREMENTS FOR SAFETY		1
Chapter 11	Scope	2
Chapter 12	Object and effects	3
Chapter 13	Fundamental requirements for safety	5
PART 2 — DEFINITIONS		7
PART 3 — ASSESSMENT OF GENERAL CHARACTERISTICS		19
Chapter 31	Purpose, supplies and structure	20
Chapter 32	External influences	22
Chapter 33	Compatibility	22
Chapter 34	Maintainability	22
PART 4 — PROTECTION FOR SAFETY		23
Chapter 41	Protection against electric shock	27
Chapter 42	Protection against thermal effects	42
Chapter 43	Protection against overcurrent	44
Chapter 44	*(Reserved for future use)*	*48*
Chapter 45	Protection against undervoltage	48
Chapter 46	Isolation and switching	48
Chapter 47	Application of protective measures for safety	51
PART 5 — SELECTION AND ERECTION OF EQUIPMENT		63
Chapter 51	Common rules	66
Chapter 52	Selection and erection of wiring systems	72
Chapter 53	Switchgear	85
Chapter 54	Earthing arrangements and protective conductors	91
Chapter 55	Other equipment	101
Chapter 56	Supplies for safety services	107
PART 6	**SPECIAL INSTALLATIONS OR LOCATIONS**	110
PART 7 — INSPECTION AND TESTING		142
Chapter 71	Initial Verification	145
Chapter 72	Alterations and additions to an installation	149
Chapter 73	Periodic inspection and testing	149
Chapter 74	Certification and Reporting	150
APPENDICES		151
1	British Standards to which reference is made in the Regulations	152
2	Statutory Regulations and associated memoranda	158
3	Time/Current Characteristics of overcurrent protective devices	160
4	Current-carrying capacity and voltage drop for cables and flexible cords	171
5	Classification of external influences	240
6	Forms of Completion and Inspection Certificate	242

Editions

The following editions have been published

FIRST EDITION	entitled 'Rules and Regulations for the Prevention of Fire Risks arising from Electric Lighting'. Issued in 1882.
SECOND EDITION	Issued in 1888.
THIRD EDITION	entitled 'General Rules recommended for Wiring for the supply of Electrical Energy'. Issued in 1897.
FOURTH EDITION	Issued in 1903.
FIFTH EDITION	entitled 'Wiring Rules'. Issued in 1907.
SIXTH EDITION	Issued in 1911.
SEVENTH EDITION	Issued in 1916.
EIGHTH EDITION	entitled 'Regulations for the Electrical Equipment of Buildings'. Issued in 1924.
NINTH EDITION	Issued in 1927.
TENTH EDITION	Issued in 1934.
ELEVENTH EDITION	Issued in 1939.
ELEVENTH EDITION (REVISED)	Issued in 1943.
ELEVENTH EDITION (REVISED 1943)	Reprinted with minor Amendments, 1945.
	Supplement issued, 1946.
	Revised Section 8 issued 1948.
TWELFTH EDITION	Issued in 1950.
	Supplement issued, 1954.
THIRTEENTH EDITION	Issued in 1955.
	Reprinted 1958, 1961, 1962 and 1964.
FOURTEENTH EDITION	Issued in 1966
	Reprinted incorporating Amendments, 1968.
	Reprinted incorporating Amendments, 1969.
	Supplement on use in metric terms issued, 1969.
	Amendments issued, 1970.
	Reprinted in metric units incorporating Amendments, 1970.
	Reprinted 1972.
	Reprinted 1973.
	Amendments issued, 1974.
	Reprinted incorporating Amendments, 1974.
	Amendments issued, 1976.
	Reprinted incorporating Amendments, 1976.
FIFTEENTH EDITION	entitled 'Regulations for Electrical Installations'. Issued in 1981. (Red Cover)
	Amendments issued, 1 January 1983.
	Reprinted incorporating Amendments, 1983. (Green Cover)
	Amendments issued, 1 May 1984.
	Reprinted incorporating Amendments, 1984. (Yellow Cover)
	Amendments issued, 1 January 1985.
	Amendments issued, 1 January 1986.
	Reprinted incorporating Amendments, 1986. (Blue Cover)
	Amendments issued, 12 June 1987.
	Reprinted incorporating Amendments, 1987. (Brown Cover)
	Reprinted with minor corrections, 1988. (Brown Cover)
SIXTEENTH EDITION	Issued in 1991. (Red Cover)
	Reprinted with minor corrections, 1992. (Red Cover)
	Reprinted as BS 7671: 1992. (Red Cover)

Wiring Regulations Committee

CONSTITUTION
as at March 1991

The President *(ex officio)*

M N John BSc FEng FIEE (Chairman)

D W M Latimer MA CEng MIEE (Vice-Chairman)

P Bingley CEng FIEE   C D Kinloch DFH CEng MIEE
J Feenan MBE CEng FIMechE FIEE   R G Parr BSc(Eng) CEng MIEE
E J B Garnham CEng MIEE   J F Wilson MBE AMIEE
P C Hoare MBE CEng FIEE   J R Worsley CEng MIEE

and the Chairmen of the Power Division Professional Groups P4 — Industrial applications and processes, and P5 — Non-industrial utilisation and electrical building services (ex officio).

and

J Wray CEng MIEE (Associated British Ports)
C Howie (Associated Offices Technical Committee)
J R Briggs BSc(Eng) CEng MIMechE MIEE FCIBSE (Association of Consulting Engineers)
D P Dossett BSc CEng MIEE (Association of Manufacturers of Domestic Electric Appliances)
F W Price CEng MIEE (Association of Supervisory and Executive Engineers)
C K Reed
D W Pryer CEng MIEE } (British Cable Makers' Confederation)
P G Newbery BSc CEng MIEE
Eur Ing J Rickwood BSc(Eng) CEng FIEE } (BEAMA Ltd)
E S Entwisle (British Coal)
R J Simpson BSc(Eng) (British Electrotechnical Approvals Board)
I M Lucking (British Radio Equipment Manufacturers' Association)
A P Storey BEng CEng MIEE (British Railways Board)
I H Campbell MSc CEng MRAeS (British Standards Institution)
J Gerrard BSc (British Telecom)
P R Smith (Bureau of Engineering Surveyors)
Eur Ing C Bennett CEng FIEE FCIBSE (Chartered Institution of Building Services Engineers)
H R Lovegrove IEng MIEIE (City & Guilds of London Institute)
A M G Minto BSc CEng MIEE (Department of the Environment)
J G Lindsay CEng FIEE (Department of Energy)
A J Lloyd CEng MIEE (Department of Trade & Industry)
C P Webber BTech CEng MIEE (Electrical Contractors' Association)
Sir James Morrison-Low Bt DFH CEng MIEE (Electrical Contractors' Association of Scotland)
J M Adams BSc(Eng) CEng FIEE (Electrical, Electronic, Telecommunications & Plumbing Union)
D W Cowe BSc CEng FIEE
M F Kennedy BSc CEng FIEE
D J Start BSc CEng MIEE } (Electricity Association Limited)
A C Wray BSc CEng MIEE
C L Astin BSc(Eng) (Energy Industries Council)
J Richman (Engineering Equipment & Material Users' Association)
M W Coates BEng (ERA Technology Ltd)
S S J Robertson DipEE CEng FIEE (Health & Safety Executive)
A J Gatfield (Home Office Fire Inspectorate)
A G Smith IEng FIEIE (Institution of Electronics & Electrical Incorporated Engineers)
Eur Ing I F Davies BSc CEng FIEE (Lighting Industry Federation)
J R Gates (London Underground Ltd)
G A Poole BSc(Eng) (Loss Prevention Council)
Major R W Dixon RE BSc(Eng) AMIEE (Ministry of Defence (Army))
Eur Ing J Stockting BSc DipEl CEng FIEE MRAeS (National Inspection Council for Electrical Installation Contracting)
K C Thompson BEng (Office of Electricity Regulation)
Appointment Pending (Royal Institute of British Architects)
J C McLuckie BSc(Eng) CEng MIEE (Scottish Office Building Directorate)
J C Owens CEng FIEE (ScottishPower)

Preface

This edition was issued on 10 May 1991. It is intended to supersede the Fifteenth Edition with effect from 1 January 1993, and until that date the Fifteenth Edition retains its validity.

The Regulations for Electrical Installations are referred to in the Electricity Supply Regulations 1988 (as amended) and the reference to the Fifteenth Edition will be changed to the Sixteenth Edition to take effect from 31 December 1992.

Installations which are scheduled for completion after 31 December 1992 should be designed in accordance with the 16th Edition.

It should be noted, however, that installations designed and constructed to the 16th Edition will not be in contravention of the 15th Edition. In particular those responsible for the design and construction of installations should make use of the 16th Edition before the 31 December 1992 in respect of certain types of installation not included in the 15th Edition, e.g. those installations described in Part 6 of the 16th Edition.

In this edition account has been taken of the technical substance of the parts of IEC Publication 364 so far published, and of the corresponding agreements reached in CENELEC. In particular, this edition takes account of the following CENELEC Harmonization Documents:

CENELEC Harmonization Document Reference		Part of the Regulation
HD 193	— Voltage Bands	PART 1 and Definitions
HD 308	— Identification and use of cores of flexible cables	PART 5
HD 384.1	— Scope	PART 1
HD 384.2	— Definitions	PART 2
HD 384.3	— Assessment of general characteristics	PART 3
HD 384.4.41	— Protection against electric shock	PART 4, CHAPTER 41
HD 384.4.42	— Protection against thermal effects	PART 4, CHAPTER 42
HD 384.4.43	— Protection against overcurrent	PART 4, CHAPTER 43
HD 384.4.45	— Protection against undervoltage	PART 4, CHAPTER 45
HD 384.4.46	— Isolation and Switching	PART 4, CHAPTER 46
HD 384.4.47	— Application of measures for Protection against electric shock	PART 4, CHAPTER 47
HD 384.4.473	— Application of measures for Protection against overcurrent	PART 4, CHAPTER 47-473
HD 384.5.51	— Selection and erection of equipment, common rules	PART 5, CHAPTER 51
HD 384.5.523	— Current-carrying capacities	PART 5, CHAPTER 52 and APPENDIX 4
HD 384.5.537	— Selection and erection of devices for isolation and switching	PART 5, CHAPTER 53-537
HD 384.5.54	— Earthing arrangements and protective conductors	PART 5, CHAPTER 54
HD 384.5.56	— Supplies for Safety services	PART 5, CHAPTER 56

In addition, account has been taken of the following CENELEC Harmonization Documents which have been approved but have not been published at the time of publication of the Sixteenth Edition, 1991:

CHAPTER 52	— Selection and Erection of Wiring Systems	CHAPTER 52
Section 702	— Swimming Pools	SECTION 602
Section 703	— Locations containing a hot air sauna heater	SECTION 603
Section 705	— Agricultural and horticultural premises	SECTION 605
Section 706	— Restrictive Conducting Locations	SECTION 606

Where the CENELEC work is still in the course of preparation the corresponding parts of this edition are based on the IEC documents.

The Regulations will be amended from time to time to take account of further progress of the international work and other developments, the arrangement of parts, chapters, and sections being intended to facilitate this.

The opportunity has also been taken to revise certain regulations for greater clarity or to take account of technical developments.

Considerable reference is made throughout the Regulations to publications of the British Standards Institution, both specifications and codes of practice. Appendix 1 lists these publications and gives their full titles whereas throughout the Regulations they are referred to only by their numbers.

Where reference is made to a British Standard in the Regulations, and the British Standard concerned takes account of a CENELEC Harmonization Document, it is understood that the reference is to be read as relating also to any foreign standard similarly based on that Harmonization Document, provided it is verified that any differences between the two standards would not result in a lesser degree of safety than that achieved by compliance with the British Standard (see Section 511).

A similar verification should be made in the case of a foreign standard based on an IEC standard but as national differences are not required to be listed in such standards, special care should be exercised (see Section 511).

Note by the Health and Safety Executive

The Health and Safety Executive (HSE) welcomes the publication of the 16th Edition of the IEE Wiring Regulations, previous Editions of which have been extensively referred to in HSE guidance over the years. The HSE continues to regard compliance with the Wiring Regulations as likely to achieve compliance with the relevant aspects of the Electricity at Work Regulations 1989. Installations may have been installed to earlier Editions of the Wiring Regulations but this does not necessarily mean that they would fail to comply with the Electricity at Work Regulations 1989.

Notes on the plan of the 16th edition

This edition is based on the plan agreed internationally for the arrangement of safety rules for electrical installations.

In the numbering system used, the first digit signifies a Part, the second digit a Chapter, the third digit a Section, and the subsequent digits the regulation number. For example, the Section number 413 is made up as follows:

PART 4 — PROTECTION FOR SAFETY
Chapter 41 (first chapter of Part 4) — Protection against electric shock.
Section 413 (third section of Chapter 41) — Protection against indirect contact.

Part 1 sets out fundamental requirements for safety that are applicable to all installations.

Part 2 defines the sense in which certain terms are used throughout the Regulations.

The subjects of the subsequent parts are as indicated below:

Part No.	Subject
3	Identification of the characteristics of the installation that will need to be taken into account in choosing and applying the requirements of the subsequent Parts. These characteristics may vary from one part of an installation to another, and should be assessed for each location to be served by the installation.
4	Description of the basic measures that are available, for the protection of persons, property and livestock and against the hazards that may arise from the use of electricity. Chapters 41 to 46 each deal with a particular hazard. Chapter 47 deals in more detail with, and qualifies, the practical application of the basic protective measures, and is divided into Sections whose numbering corresponds to the numbering of the preceding chapters; thus Section 471 needs to be read in conjunction with Chapter 41, Section 473 with Chapter 43, and Section 476 with Chapter 46.
5	Precautions to be taken in the selection and erection of the equipment of the installation. Chapter 51 relates to equipment generally and Chapters 52 to 56 to particular types of equipment.
6	Special installations or locations — particular requirements.
7	Inspection and testing.

The sequence of the plan should be followed in considering the application of any particular requirement of the Regulations. The general index provides a ready reference to particular regulations by subject, but in applying any one regulation the requirements of related regulations should be borne in mind. Cross-references are provided, and the index is arranged, to facilitate this.

Individual Regulations are identified by a three-part number, e.g. 413-02-27 and thus 16th Edition Regulations are distinguished from those of the 15th Edition.

In many cases a group of associated Regulations is covered by a side heading which is identified by a two-part number, e.g. 547-03. Throughout the Regulations where reference is made to such a two-part number, that reference is to be taken to include all the individual Regulation numbers which are covered by that side heading and include that two-part number.

PART 1

SCOPE, OBJECT AND FUNDAMENTAL REQUIREMENTS FOR SAFETY

CONTENTS

CHAPTER 11	SCOPE	
110-01	General	
110-02	Exclusions from scope	
110-03	Voltage ranges	
110-04	Equipment	
CHAPTER 12	OBJECT AND EFFECTS	
120-01	General	
120-02	Relationship with Statutory Regulations	
120-03	Installations in premises subject to licensing	
120-04	Use of established materials, equipment and methods	
120-05	New materials and inventions	
CHAPTER 13	FUNDAMENTAL REQUIREMENTS FOR SAFETY	
130-01	Workmanship and materials	
130-02	General	
130-03	Overcurrent protective devices	
130-04	Precautions against earth leakage and earth fault currents	
130-05	Protective devices and switches	
130-06	Isolation and switching	
130-07	Accessibility of equipment	
130-08	Precautions in adverse conditions	
130-09	Additions and alterations to an installation	
130-10	Inspection and testing	

PART 1

Scope, Object and Fundamental Requirements for Safety

CHAPTER 11

SCOPE

110-01 General

110-01-01 The Regulations apply to the design, selection, erection, inspection and testing of electrical installations, other than those excluded by Regulation 110-02.

Particular requirements are included for electrical installations of:

- (i) locations containing a bath tub or shower unit
- (ii) swimming pools
- (iii) locations containing a hot air sauna
- (iv) construction sites
- (v) agricultural and horticultural premises
- (vi) restrictive conducting locations
- (vii) caravans and caravan parks
- (viii) highway power supplies and street furniture.

The Regulations are intended to be applied to electrical installations generally but, in certain cases, they may need to be supplemented by the requirements or recommendations of British Standards or by the requirements of the person ordering the work.

Such cases include the following:

- (ix) electric signs and high voltage luminous discharge tube installations - BS 559
- (x) emergency lighting - BS 5266
- (xi) installations in potentially explosive atmospheres - BS 5345
- (xii) fire detection and alarm systems in buildings - BS 5839
- (xiii) installations subject to the Telecommunications Act 1984 - BS 6701, Part 1
- (xiv) electric surface heating systems - BS 6351.

110-02 Exclusions from Scope

110-02-01 The Regulations do not apply to the following installations:

- (i) 'supplier's works' as defined in The Electricity Supply Regulations, 1988, as amended
- (ii) railway traction equipment, rolling stock and signalling equipment
- (iii) equipment of motor vehicles, except those to which the requirements of the Regulations concerning caravans are applicable
- (iv) equipment on board ships
- (v) equipment of mobile and fixed offshore installations
- (vi) equipment of aircraft
- (vii) those aspects of mines and quarries specifically covered by Statutory Regulations
- (viii) radio interference suppression equipment, except so far as it affects safety of the electrical installation
- (ix) lightning protection of buildings, except in respect of the equipotential bonding connection to the electrical installation, BS 6651
- (x) those aspects of lift installations covered by BS 5655.

110-03 Voltage Ranges

110-03-01 Installations utilising the following nominal voltage ranges are dealt with:

(i) extra-low voltage
(ii) low voltage.

Regulations are also included for certain installations operating at voltages exceeding low voltage.

110-04 Equipment

110-04-01 The Regulations apply to items of electrical equipment only so far as selection and application of the equipment in the installation are concerned. The Regulations do not deal with requirements for the construction of prefabricated assemblies of electrical equipment, which are required to comply with appropriate specifications.

CHAPTER 12

OBJECT AND EFFECTS

120-01 General

120-01-01 The Regulations are designed to protect:

(i) persons
(ii) property and
(iii) livestock, in locations intended for them,
against hazards arising from an electrical installation used with reasonable care having regard to the purpose for which the installation is intended.

The requirements relate to protection against:

(iv) electric shock
(v) fire
(vi) burns
(vii) injury from mechanical movement of electrically actuated equipment, in so far as such injury is intended to be prevented by electrical emergency switching or by electrical switching for mechanical maintenance of non-electrical parts of such equipment.

120-01-02 The Regulations are intended to be cited in their entirety if referred to in any contract. They are not intended to take the place of a detailed specification or to instruct untrained persons or to provide for every circumstance. Installations of a difficult or special character will require the advice of a suitably qualified electrical engineer.

120-02 Relationship with Statutory Regulations

120-02-01 The Regulations are non-statutory regulations. They may, however, be used in a court of law in evidence to claim compliance with a statutory requirement. The relevant statutory provisions are listed in Appendix 2 and include Acts of Parliament and Regulations made thereunder. In some cases Regulations may be accompanied by Codes of Practice approved under Section 16 of the Health and Safety at Work Act, 1974. The legal status of these Codes is explained in Section 17 of the 1974 Act.

For a supply given in accordance with the Electricity Supply Regulations 1988, as amended, it shall be deemed that the connection with earth of the neutral of the supply is permanent. Outside Great Britain, confirmation shall be sought from the supplier that the supply conforms to requirements corresponding to those of the Electricity Supply Regulations 1988, as amended, in this respect.

Parts 3 to 7 of the Regulations set out in greater detail methods and practices which are regarded as meeting the requirements of Chapter 13. Any departure from those Parts requires special consideration by the designer of the installation and is to be noted in the Completion Certificate specified in Part 7.

120-03 Installations in Premises subject to Licensing

120-03-01 For installations in premises over which a licensing or other authority exercises a statutory control, the requirements of that authority are to be ascertained and complied with in the design and execution of the installation.

120-04 Use of Established Materials, Equipment and Methods

120-04-01 Only established materials, equipment and methods are considered, but it is not intended to discourage invention or to exclude other materials, equipment and methods affording an equivalent degree of safety which may be developed in the future.

120-05 New Materials and Inventions

120-05-01 Where the use of a new material or invention leads to departures from the Regulations, the resulting degree of safety of the installation is to be not less than that obtained by compliance with the Regulations.

Such use is to be noted on the Completion Certificate specified in Part 7.

CHAPTER 13

FUNDAMENTAL REQUIREMENTS FOR SAFETY

130-01 **Workmanship and Materials**

130-01-01 Good workmanship and proper materials shall be used.

130-02 **General**

130-02-01 All equipment shall be constructed, installed and protected and shall be capable of being maintained, inspected and tested, so as to prevent danger so far as is reasonably practicable.

130-02-02 All equipment shall be suitable for the maximum power demanded by the current-using equipment when it is functioning in its intended manner.

130-02-03 All electrical conductors shall be of sufficient size and current-carrying capacity for the purposes for which they are intended.

130-02-04 All conductors shall either:

(i) be so insulated and where necessary further effectively protected, or

(ii) be so placed and safeguarded as to prevent danger, so far as is reasonably practicable.

130-02-05 Every electrical joint and connection shall be of proper construction as regards conductance, insulation, mechanical strength and protection.

130-03 **Overcurrent Protective Devices**

130-03-01 Where necessary to prevent danger, every installation and every circuit thereof shall be protected against overcurrent by devices which:

(i) will operate automatically at values of current which are suitably related to the safe current rating of the circuit, and
(ii) are of adequate breaking capacity and where appropriate, making capacity, and
(iii) are suitably located and are constructed so as to prevent danger from overheating, arcing or the scattering of hot particles when they come into operation and to permit ready restoration of the supply without danger.

130-04 **Precautions against Earth Leakage and Earth Fault Currents**

130-04-01 Where metalwork of electrical equipment, other than current-carrying conductors, may become charged with electricity in such a manner as to cause danger:

(i) the metalwork shall be connected with earth in such a manner as will cause discharge of electrical energy without danger, or

(ii) other equally effective precautions shall be taken to prevent danger.

130-04-02 Every circuit shall be arranged so as to prevent the persistence of dangerous earth leakage currents.

130-04-03 Where metalwork is connected with earth in accordance with Regulation 130-04-01(i) the circuits concerned shall be protected against the persistence of an earth fault current by:

(i) the overcurrent protective devices required by Regulation 130-03-01, or

(ii) a residual current device or equally effective device.

The method described in item (ii) above shall be used whenever the prospective earth fault current is insufficient to cause prompt operation of the overcurrent protective devices.

130-04-04 Where any metalwork of electrical equipment is connected with earth in accordance with Regulation 130-04-01(i) and is accessible simultaneously with substantial exposed metal parts of other services, the latter shall be effectively connected to the main earthing terminal of the installation.

130-05 Protective Devices and Switches

130-05-01 A single-pole fuse, switch or circuit-breaker shall be inserted in the phase conductor only.

130-05-02 No switch or circuit-breaker, excepting where linked, or fuse, shall be inserted in an earthed neutral conductor and any linked switch or linked circuit-breaker inserted in an earthed neutral conductor shall be arranged to break all the related phase conductors.

130-06 Isolation and Switching

130-06-01 Effective means, suitably placed for ready operation, shall be provided so that all voltage may be cut off from every installation, from every circuit thereof and from all equipment, as may be necessary to prevent or remove danger.

130-06-02 Every fixed electric motor shall be provided with an efficient means of switching off, readily accessible, easily operated and so placed as to prevent danger.

130-07 Accessibility of Equipment

130-07-01 Every piece of equipment which requires operation or attention by a person shall be so installed that adequate and safe means of access and working space are afforded for such operation or attention.

130-08 Precautions in Adverse Conditions

130-08-01 All equipment likely to be exposed to weather, corrosive atmospheres or other adverse conditions, shall be so constructed or protected as may be necessary to prevent danger arising from such exposure.

130-08-02 All equipment in surroundings susceptible to risk of fire or explosion shall be so constructed or protected and such other special precautions shall be taken, as may be necessary to prevent danger.

130-09 Additions and Alterations to an Installation

130-09-01 No addition or alteration, temporary or permanent, shall be made to an existing installation, unless it has been ascertained that the rating and the condition of any existing equipment, including that of the supplier, which will have to carry any additional load is adequate for the altered circumstances and the earthing arrangement is also adequate.

130-10 Inspection and Testing

130-10-01 On completion of an installation or an extension or alteration of an installation, appropriate tests and inspection shall be made, to verify so far as is reasonably practicable that the requirements of Regulations 130-01 to 130-09 have been met. The person carrying out the test and the inspection, or a person acting on his behalf, shall inform the person ordering the work of the recommendations for Periodic Inspection and Testing described in Chapter 73.

PART 2

DEFINITIONS

For the purposes of the Regulations the following definitions shall apply. Some of these definitions are aligned with those given in BS 4727 - 'Glossary of electrotechnical, power, telecommunication, electronics, lighting and colour terms'. Other terms not defined herein are used in the sense defined in BS 4727.

Accessory. A device, other than current-using equipment, associated with such equipment or with the wiring of an installation.

Ambient temperature. The temperature of the air or other medium where the equipment is to be used.

Appliance. An item of current-using equipment other than a luminaire or an independent motor.

Arm's reach. A zone of accessibility to touch, extending from any point on a surface where persons usually stand or move about, to the limits which a person can reach with his hand in any direction without assistance.

This zone of accessibility is illustrated by Figure 1 in which the values refer to bare hands without any assistance, e.g. from tools or a ladder.

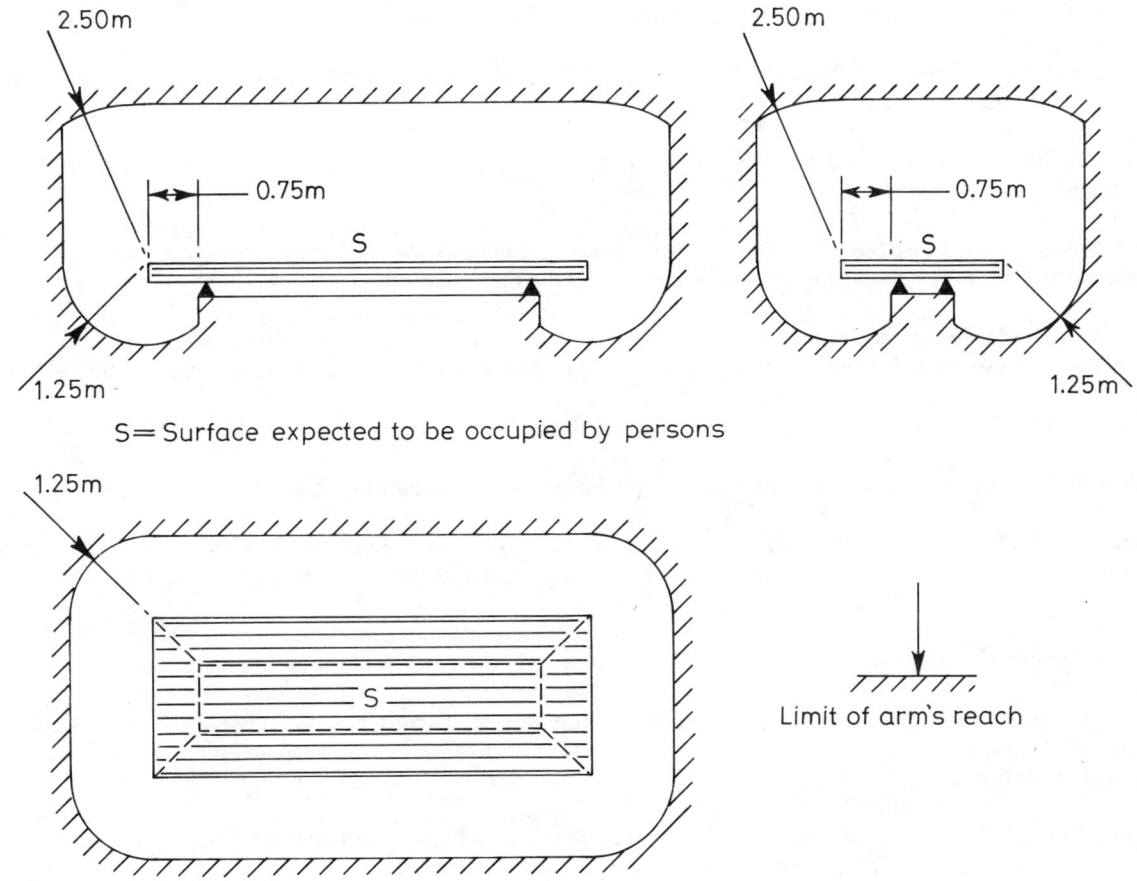

Fig. 1

Barrier. A part providing a defined degree of protection against contact with live parts, from any usual direction of access.

Basic insulation. Insulation applied to live parts to provide basic protection against electric shock and which does not necessarily include insulation used exclusively for functional purposes.

Bonding conductor. A protective conductor providing equipotential bonding.

Building void, accessible. A space within the structure or the components of a building accessible only at certain points.

Building void, non-accessible. A space within a structure or the components of a building which has no ready means of access.

Bunched. Cables are said to be bunched when two or more are contained within a single conduit, duct, ducting, or trunking or, if not enclosed, are not separated from each other by a specified distance.

Cable bracket. A horizontal cable support system, consisting of elements fixed at one end only, spaced at intervals along the length of the cable and on which the cable rests.

Cable channel. An enclosure situated above or in the ground, ventilated or closed, and having dimensions which do not permit the access of persons but allow access to the conductors and/or cables throughout their length during and after installation. A cable channel may or may not form part of the building construction.

Cable Cleat. A component of a support system, which consists of elements spaced at intervals along the length of the cable or conduit and which mechanically retains the cable or conduit.

Cable coupler. A means of enabling the connection or disconnection, at will, of two flexible cables. It consists of a connector and a plug.

Cable ducting. A manufactured enclosure of metal or insulating material, other than conduit or cable trunking, intended for the protection of cables which are drawn-in after erection of the ducting.

Cable ladder. A cable support consisting of a series of supporting elements rigidly fixed to main supporting members. The supporting elements occupy less than 10% of the plan area.

Cable tray. A cable support consisting of a continuous base with raised edges and no covering. A cable tray is considered to be non-perforated, where less than 30% of the material is removed from the base.

Cable trunking. A manufactured enclosure for the protection of cables, normally of rectangular cross-section, of which one side is removable or hinged.

Cable tunnel. An enclosure (corridor) containing supporting structures for conductors and/or cables and joints and whose dimensions allow persons to pass freely throughout the entire length.

Caravan. A trailer leisure accommodation vehicle, used for touring, designed to meet the requirements for the construction and use of road vehicles. (See also definition of Motor caravan and Leisure accommodation vehicle.)

Caravan Park. An area of land that contains two or more caravan pitches.

Caravan Pitch. A plot of ground upon which a single leisure accommodation vehicle or leisure home may stand.

Caravan pitch electrical supply equipment. Equipment that provides means of connecting and disconnecting supply cables from a leisure accommodation vehicle to a fixed external power supply.

Cartridge fuse link. A device comprising a fuse element or several fuse elements connected in parallel enclosed in a cartridge usually filled with arc-extinguishing medium and connected to terminations. See fuse link.

Circuit. An assembly of electrical equipment supplied from the same origin and protected against overcurrent by the same protective device(s). For the purposes of Chapter 52 of these Regulations, certain types of circuit are categorised as follows:

Category 1 circuit - A circuit (other than a fire alarm or emergency lighting circuit) operating at low voltage and supplied directly from a mains supply system.

Category 2 circuit - With the exception of fire alarm and emergency lighting circuits, any circuit for telecommunication (e.g. radio, telephone, sound distribution, intruder alarm, bell and call, and data transmission circuit) which is supplied at extra-low voltage.

Category 3 circuit - A fire alarm circuit or an emergency lighting circuit.

Circuit-breaker. A device capable of making, carrying and breaking normal load currents and also making and automatically breaking, under pre-determined conditions, abnormal currents such as short-circuit currents. It is usually required to operate infrequently although some types are suitable for frequent operation.

Circuit-breaker (linked). A circuit-breaker the contacts of which are so arranged as to make or break all poles simultaneously or in a definite sequence.

Circuit protective conductor (cpc). A protective conductor connecting exposed-conductive-parts of equipment to the main earthing terminal.

Class I equipment. Equipment in which protection against electric shock does not rely on basic insulation only, but which includes means for the connection of exposed-conductive-parts to a protective conductor in the fixed wiring of the installation (see BS 2754).

Class II equipment. Equipment in which protection against electric shock does not rely on basic insulation only, but in which additional safety precautions such as supplementary insulation are provided, there being no provision for the connection of exposed metal work of the equipment to a protective conductor, and no reliance upon precautions to be taken in the fixed wiring of the installation (see BS 2754).

Conduit. A part of a closed wiring system for cables in electrical installations, allowing them to be drawn in and/or replaced, but not inserted laterally.

Connector. The part of a cable coupler or of an appliance coupler which is provided with female contacts and is intended to be attached to the end of the flexible cable remote from the supply.

Current-carrying capacity of a conductor. The maximum current which can be carried by a conductor under specified conditions without its steady state temperature exceeding a specified value.

Current-using equipment. Equipment which converts electrical energy into another form of energy, such as light, heat, or motive power.

Danger. Risk of injury to persons (and livestock where expected to be present) from:

(i) fire, electric shock, and burns arising from the use of electrical energy, and

(ii) mechanical movement of electrically controlled equipment, in so far as such danger is intended to be prevented by electrical emergency switching or by electrical switching for mechanical maintenance of non-electrical parts of such equipment.

Data processing equipment. Electrically operated machine units that, separately or assembled in systems, accumulate, process and store data. Acceptance and divulgence of data may or may not be by electronic means.

Design current (of a circuit). The magnitude of the current (r.m.s. value for a.c.) to be carried by the circuit in normal service.

Direct contact. Contact of persons or livestock with live parts which may result in electric shock.

Distribution board. An assembly containing switching or protective devices (e.g. fuses or circuit-breakers) associated with one or more outgoing circuits fed from one or more incoming circuits, together with terminals for the neutral and protective circuit conductors. It may also include signalling and other control devices. Means of isolation may be included in the board or may be provided separately.

Distribution circuit. A Category 1 circuit connecting the origin of the installation to:

(i) an item of switchgear, or
(ii) an item of controlgear, or
(iii) a distribution board

to which one or more final circuits or items of current-using equipment are connected. (See also definition of Final circuit.)

A distribution circuit may also connect the origin of an installation to an outlying building or separate installation, when it is sometimes called a sub-main.

Double insulation. Insulation comprising both basic insulation and supplementary insulation.

Duct. A closed passageway formed underground or in a structure and intended to receive one or more cables which may be drawn in.

Ducting. *(See Cable Ducting)*

Earth. The conductive mass of the Earth, whose electric potential at any point is conventionally taken as zero.

Earth electrode. A conductor or group of conductors in intimate contact with, and providing an electrical connection to, Earth.

Earth electrode resistance. The resistance of an earth electrode to Earth.

Earth fault current. A fault current which flows to Earth.

Earth fault loop impedance. The impedance of the earth fault current loop starting and ending at the point of earth fault. This impedance is denoted by the symbol Z_S.

The earth fault loop comprises the following, starting at the point of fault:

- the circuit protective conductor, and
- the consumer's earthing terminal and earthing conductor, and
- for TN systems, the metallic return path, and
- for TT and IT systems, the earth return path, and
- the path through the earthed neutral point of the transformer
  and the transformer winding, and
- the phase conductor from the transformer to the point of fault.

Earth leakage current. A current which flows to Earth, or to extraneous-conductive-parts, in a circuit which is electrically sound. This current may have a capacitive component including that resulting from the deliberate use of capacitors.

Earthed concentric wiring. A wiring system in which one or more insulated conductors are completely surrounded throughout their length by a conductor, for example a metallic sheath, which acts as a PEN conductor.

Earthed equipotential zone. A zone within which exposed-conductive-parts and extraneous-conductive-parts are maintained at substantially the same potential by bonding, such that, under fault conditions, the differences in potential between simultaneously accessible exposed and extraneous-conductive-parts will not cause electric shock.

Earthing. The act of connecting the exposed-conductive-parts of an installation to the main earthing terminal of an installation.

Earthing conductor. A protective conductor connecting the main earthing terminal of an installation to an earth electrode or to other means of earthing.

Electric shock. A dangerous physiological effect resulting from the passing of an electric current through a human body or livestock.

Electrical equipment (abbr: *Equipment*). Any item for such purposes as generation, conversion, transmission, distribution or utilisation of electrical energy, such as machines, transformers, apparatus, measuring instruments, protective devices, wiring materials, accessories, appliances and luminaires.

Electrical installation. (abbr: *Installation*). An assembly of associated electrical equipment supplied from a common origin to fulfil a specific purpose and having certain co-ordinated characteristics.

Electrically independent earth electrodes. Earth electrodes located at such a distance from one another that the maximum current likely to flow through one of them does not significantly affect the potential of the other(s).

Electrode boiler (or electrode water heater). Equipment for the electrical heating of water or electrolyte by the passage of an electric current between electrodes immersed in the water or electrolyte.

Emergency stopping. Emergency switching intended to stop a dangerous movement.

Emergency switching. Rapid cutting off of electrical energy to remove any unexpected hazard to persons, livestock, or property.

Enclosure. A part providing an appropriate degree of protection of equipment against certain external influences and a defined degree of protection against contact with live parts from any direction.

Equipment. (abbr: *see Electrical equipment*).

Equipotential bonding. Electrical connection maintaining various exposed-conductive-parts and extraneous-conductive-parts at substantially the same potential.

Equipotential zone. (*see Earthed equipotential zone*).

Exposed-conductive-part. A conductive part of equipment which can be touched and which is not a live part but which may become live under fault conditions.

External influence. Any influence external to an electrical installation which affects the design and safe operation of that installation.

Extraneous-conductive-part. A conductive part liable to introduce a potential, generally earth potential, and not forming part of the electrical installation.

Fault. A circuit condition in which current flows through an abnormal or unintended path. This may result from an insulation failure or the bridging of insulation. Conventionally the impedance between live conductors or between live conductors and exposed or extraneous-conductive-parts at the fault position is considered negligible.

Fault current. A current resulting from a fault.

Final circuit. A circuit connected directly to current-using equipment, or to a socket-outlet or socket-outlets or other outlet points for the connection of such equipment.

Fixed equipment. Equipment fastened to a support or otherwise secured in a specific location.

Flexible wiring system. A wiring system designed to provide mechanical flexibility in use without degradation of the electrical components.

Functional earthing. Connection to Earth necessary for proper functioning of electrical equipment.

Functional extra-low voltage. Any Extra-low voltage system in which not all of the protective measures required for SELV have been applied.

Fuse. A device that by the fusing of one or more of its specially designed and proportioned components, opens the circuit in which it is inserted by breaking the current when this exceeds a given value for a sufficient time. The fuse comprises all the parts that form the complete device.

Fuse element. A part of a fuse designed to melt when the fuse operates.

Fuse link. A part of a fuse, including the fuse element(s), which requires replacement by a new or renewable fuse link after the fuse has operated and before the fuse is put back into service.

Gas installation pipe. Any pipe, not being a service pipe (other than any part of a service pipe comprised in a primary meter installation) or a pipe comprised in a gas appliance, for conveying gas for a particular consumer and including any associated valve or other gas fitting.

Highway. A Highway means any way (other than a waterway) over which there is public passage and includes the Highway verge and any bridge over which, or tunnel through which, the Highway passes.

Highway distribution board. A fixed structure or underground chamber, located on a Highway, used as a distribution point, for connecting more than one Highway distribution circuit to a common origin. Street furniture which supplies more than one circuit is defined as a Highway distribution board. The connection of a single temporary load to an item of Street furniture shall not in itself make that item of Street furniture into a Highway distribution board.

Highway distribution circuit. A category 1 circuit connecting the origin of the installation to a remote Highway distribution board or items of Street furniture. It may also connect a Highway distribution board to Street furniture.

Highway power supply. An electrical installation comprising an assembly of associated Highway distribution circuits, Highway distribution boards and Street furniture, supplied from a common origin.

Hot air sauna. A room or location in which air is heated to a high temperature and in which the relative humidity is normally low, rising for short periods of time only when water is poured over the heater.

Indirect contact. Contact of persons or livestock with exposed-conductive-parts made live by a fault and which may result in electric shock.

Installation. (abbr: *see Electrical installation*).

Instructed person. A person adequately advised or supervised by skilled persons to enable him to avoid dangers which electricity may create.

Insulation. Suitable non-conductive material enclosing, surrounding, or supporting a conductor.

Isolation. A function intended to cut off for reasons of safety the supply from all, or a discrete section, of the installation by separating the installation or section from every source of electrical energy.

Isolator. A mechanical switching device which provides the function of isolation.

Ladder. (*See Cable ladder*).

Leisure accommodation vehicle. Unit of living accommodation for temporary or seasonal occupation which may meet requirements for construction and use of road vehicles.

Live part. A conductor or conductive part intended to be energised in normal use, including a neutral conductor but, by convention, not a PEN conductor.

Low noise earth. An earth connection in which the level of conducted or induced interference from external sources does not produce an unacceptable incidence of malfunction in the data processing or similar equipment to which it is connected. The susceptibility in terms of amplitude/frequency characteristics varies depending on the type of equipment.

Luminaire. Equipment which distributes, filters, or transforms the light from one or more lamps, and which includes any parts necessary for supporting, fixing and protecting the lamps, but not the lamps themselves, and, where necessary, circuit auxiliaries together with the means for connecting them to the supply. For the purposes of the Regulations a lampholder, however supported, is deemed to be a luminaire.

Luminaire supporting coupler (LSC). A means, comprising an LSC outlet and an LSC plug, providing mechanical support for a luminaire and the electrical connection to and disconnection from a fixed wiring installation.

LV switchgear and controlgear assembly. A combination of one or more low-voltage switching devices together with associated control, measuring, signalling, protective, regulating equipment, etc., completely assembled under the responsibility of the manufacturer with all the internal electrical and mechanical interconnection and structural parts. The components of the assembly may be electromechanical or electronic. The assembly may be either type-tested or partially type-tested (see BS 5486 Part 1).

Main earthing terminal. The terminal or bar provided for the connection of protective conductors, including equipotential bonding conductors, and conductors for functional earthing if any, to the means of earthing.

Mechanical maintenance. The replacement, refurbishment or cleaning of lamps and non-electrical parts of equipment, plant and machinery.

Motor caravan. Self-propelled leisure accommodation vehicle, used for touring, designed to meet requirements for the construction and use of road vehicles. The accommodation may be fixed or demountable. (See also definition of Caravan.)

Neutral conductor. A conductor connected to the neutral point of a system and contributing to the transmission of electrical energy. The term also means the equivalent conductor of an IT or d.c. system unless otherwise specified in the Regulations.

Nominal voltage. (*See Voltage, nominal*).

Obstacle. A part preventing unintentional contact with live parts but not preventing deliberate contact.

Origin of an installation. The position at which electrical energy is delivered to an electrical installation.

Overcurrent. A current exceeding the rated value. For conductors the rated value is the current-carrying capacity.

Overcurrent detection. A method of establishing that the value of current in a circuit exceeds a predetermined value for a specified length of time.

Overload current. An overcurrent occurring in a circuit which is electrically sound.

PEN conductor. A conductor combining the functions of both protective conductor and neutral conductor.

Phase conductor. A conductor of an a.c. system for the transmission of electrical energy other than a neutral conductor, a protective conductor or a PEN conductor. The term also means the equivalent conductor of a d.c. system unless otherwise specified in the Regulations.

Plug. A device, provided with contact pins, which is intended to be attached to a flexible cable, and which can be engaged with a socket-outlet or with a connector.

Point (in wiring). A termination of the fixed wiring intended for the connection of current-using equipment.

Portable equipment. Electrical equipment which is moved while in operation or which can easily be moved from one place to another while connected to the supply.

Prospective fault current. The value of overcurrent at a given point in a circuit resulting from a fault of negligible impedance between live conductors having a difference of potential under normal operating conditions, or between a live conductor and an exposed-conductive-part.

Protective conductor. A conductor used for some measures of protection against electric shock and intended for connecting together any of the following parts:

   (i)   exposed-conductive-parts
   (ii) extraneous-conductive-parts
   (iii) the main earthing terminal
   (iv) earth electrode(s)
   (v)   the earthed point of the source, or an artificial neutral.

EXAMPLE OF EARTHING ARRANGEMENTS AND PROTECTIVE CONDUCTORS
(see Chapter 54)

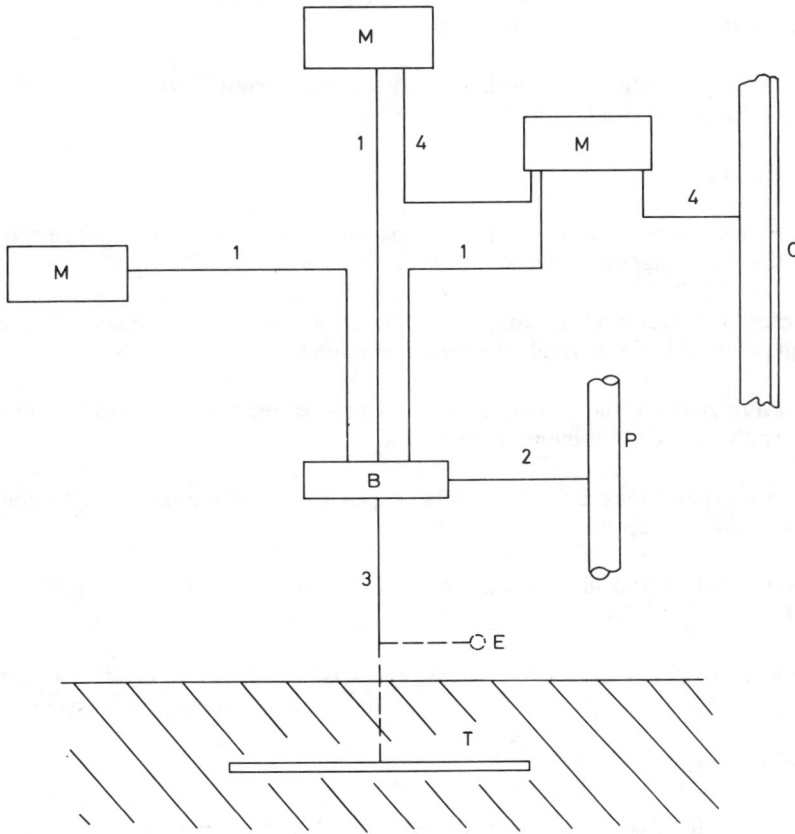

Fig. 2

1,2,3,4,	=	protective conductors
1	=	circuit protective conductor
2	=	main equipotential bonding conductor
3	=	earthing conductor
4	=	supplementary equipotential bonding conductors (where required)
B	=	main earthing terminal
M	=	exposed-conductive-part
C	=	extraneous-conductive-part
P	=	main metallic water pipe
T	=	earth electrode (TT and IT systems)
E	=	other means of earthing (TN systems)

Reduced low voltage system. A system in which the nominal phase to phase voltage does not exceed 110 volts and the nominal phase to earth voltage does not exceed 63.5 volts.

Reinforced insulation. Single insulation applied to live parts, which provides a degree of protection against electric shock equivalent to double insulation under the conditions specified in the relevant standard.
The term 'single insulation' does not imply that the insulation must be one homogeneous piece. It may comprise several layers which cannot be tested singly as supplementary or basic insulation.

Residual current. The vector sum of the instantaneous values of current flowing through all live conductors of a circuit at a point in the electrical installation.

Residual current device. A mechanical switching device or association of devices intended to cause the opening of the contacts when the residual current attains a given value under specified conditions.

Residual operating current. Residual current which causes the residual current device to operate under specified conditions.

Resistance area (for an earth electrode only). The surface area of ground (around an earth electrode) on which a significant voltage gradient may exist.

Restrictive conductive location. A location comprised mainly of metallic or conductive surrounding parts, within which it is likely that a person will come into contact through a substantial portion of his body with the conductive surrounding parts and where the possibility of preventing this contact is limited.

Ring final circuit. A final circuit arranged in the form of a ring and connected to a single point of supply.

Safety service. An electrical system for electrical equipment provided to protect or warn persons in the event of a hazard, or essential to their evacuation from a location.

SELV. An Extra-low voltage system which is electrically separated from Earth and from other systems in such a way that a single fault cannot give rise to the risk of electric shock.

Shock. (*See Electric shock*).

Shock current. A current passing through the body of a person or livestock such as to cause electric shock and having characteristics likely to cause dangerous effects.

Short-circuit current. An overcurrent resulting from a fault of negligible impedance between live conductors having a difference in potential under normal operating conditions.

Simultaneously accessible parts. Conductors or conductive parts which can be touched simultaneously by a person or, in locations specifically intended for them, by livestock.

Simultaneously accessible parts may be: live parts, exposed-conductive-parts, extraneous-conductive-parts, protective conductors, or earth electrodes.

Skilled person. A person with technical knowledge or sufficient experience to enable him to avoid dangers which electricity may create.

Socket-outlet. A device, provided with female contacts, which is intended to be installed with the fixed wiring, and intended to receive a plug. A luminaire track system is not regarded as a socket-outlet system.

Spur. A branch from a ring final circuit.

Stationary equipment. Electrical equipment which is either fixed, or equipment having a mass exceeding 18 kg and not provided with a carrying handle.

Street furniture. Fixed equipment, located on a Highway, the purpose of which is directly associated with the use of the Highway.

Street located equipment. Fixed equipment, located on a Highway, the purpose of which is not directly associated with the use of the Highway.

Supplementary insulation. Independent insulation applied in addition to basic insulation in order to provide protection against electric shock in the event of a failure of basic insulation.

Supplier. A person who supplies electrical energy, and, where electric lines and apparatus used for that purpose are owned otherwise than by that person shall include the owner of those electric lines and apparatus.

Supplier's works. Electric lines, supports and apparatus of, or under the control of, a supplier used for the purposes of supply, and cognate expressions shall be construed accordingly.

Switch. A mechanical device capable of making, carrying and breaking current under normal circuit conditions, which may include specified operating overload conditions, and also of carrying for a specified time currents under specified abnormal circuit conditions such as those of short-circuit. It may also be capable of making, but not breaking, short-circuit currents.

Switch, linked. A switch the contacts of which are so arranged as to make or break all poles simultaneously or in a definite sequence.

Switchboard. An assembly of switchgear with or without instruments, but the term does not apply to groups of local switches in final circuits.

Switchgear. An assembly of main and auxiliary switching apparatus for operation, regulation, protection or other control of an electrical installation.

System. An electrical system consisting of a single source of electrical energy and an installation. For certain purposes of the Regulations, types of system are identified as follows, depending upon the relationship of the source, and of exposed - conductive - parts of the installation, to Earth:

- **TN system,** a system having one or more points of the source of energy directly earthed, the exposed-conductive-parts of the installation being connected to that point by protective conductors,

- **TN-C system,** in which neutral and protective functions are combined in a single conductor throughout the system, (see Figure 3),

- **TN-S system**, having separate neutral and protective conductors throughout the system, (see Figure 4),

- **TN-C-S system,** in which neutral and protective functions are combined in a single conductor in part of the system, (see Figure 5),

- **TT system,** a system having one point of the source of energy directly earthed, the exposed-conductive-parts of the installation being connected to earth electrodes electrically independent of the earth electrodes of the source, (see Figure 6),

- **IT system,** a system having no direct connection between live parts and Earth, the exposed-conductive-parts of the electrical installation being earthed, (see Figure 7).

Temporary supply unit. An enclosure containing equipment for the purpose of taking a temporary electrical supply safely from an item of Street furniture.

Trunking. (*See Cable trunking*).

Voltage, nominal. Voltage by which an installation (or part of an installation) is designated. The following ranges of nominal voltage (r.m.s. values for a.c.) are defined:

- **Extra-low** Normally not exceeding 50 V a.c. or 120 V ripple free d.c., whether between conductors or to Earth,

- **Low** Normally exceeding extra-low voltage but not exceeding 1000 V a.c. or 1500 V d.c. between conductors, or 600 V a.c. or 900 V d.c. between conductors and Earth.

The actual voltage of the installation may differ from the nominal value by a quantity within normal tolerances.

Voltage, reduced. (*See Reduced low voltage system*).

Wiring system. An assembly made up of cable or busbars and parts which secure and, if necessary, enclose the cable or busbars.

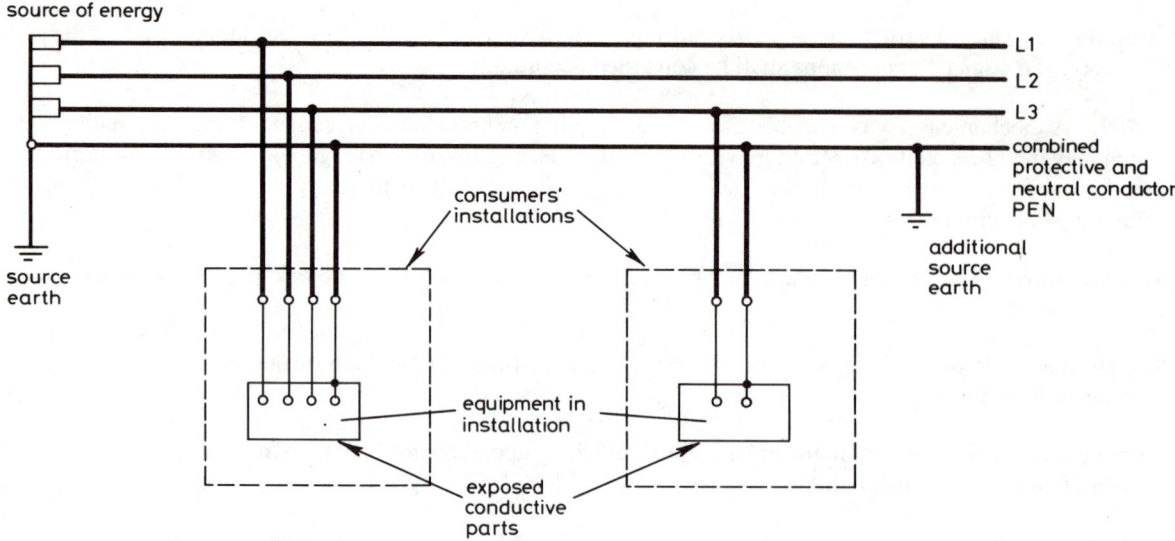

Fig. 3   TN-C system. Neutral and protective functions combined in a single conductor throughout system.

All exposed-conductive-parts of an installation are connected to the PEN conductor.

An example of the TN-C arrangement is earthed concentric wiring but where it is intended to use this special authorisation must be obtained from the appropriate authority.

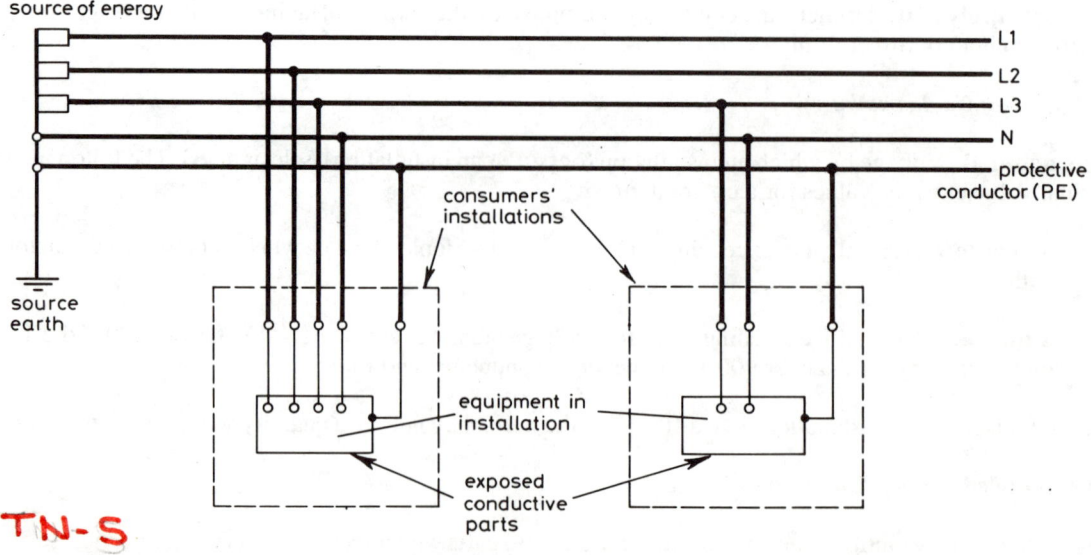

Fig. 4   TN-C system. Separate neutral and protective conductors throughout system.

The protective conductor (PE) is the metallic covering of the cable supplying the installations or separate conductor.

All exposed-conductive-parts of an installation are connected to this protective conductor via the main earthing terminal of the installation.

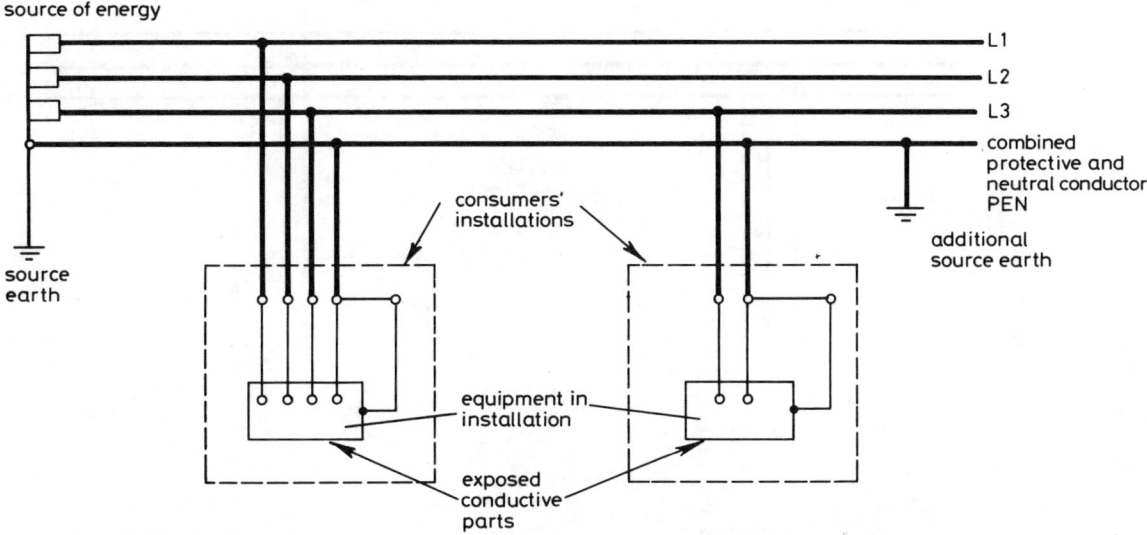

Fig. 5   TN-C-S system. Neutral and protective functions combined in a single conductor in a part of the system.

The usual form of a TN-C-S system is as shown, where the supply is TN-C and the arrangement in the installations is TN-S

This type of distribution is known also as Protective Multiple Earthing and the PEN conductor is referred to as the combined neutral and earth (CNE) conductor.

The supply system PEN conductor is earthed at several points and an earth electrode may be necessary at or near a consumer's installation.

All exposed-conductive-parts of an installation are connected to the PEN conductor via the main earthing terminal and the neutral terminal, these terminals being linked together.

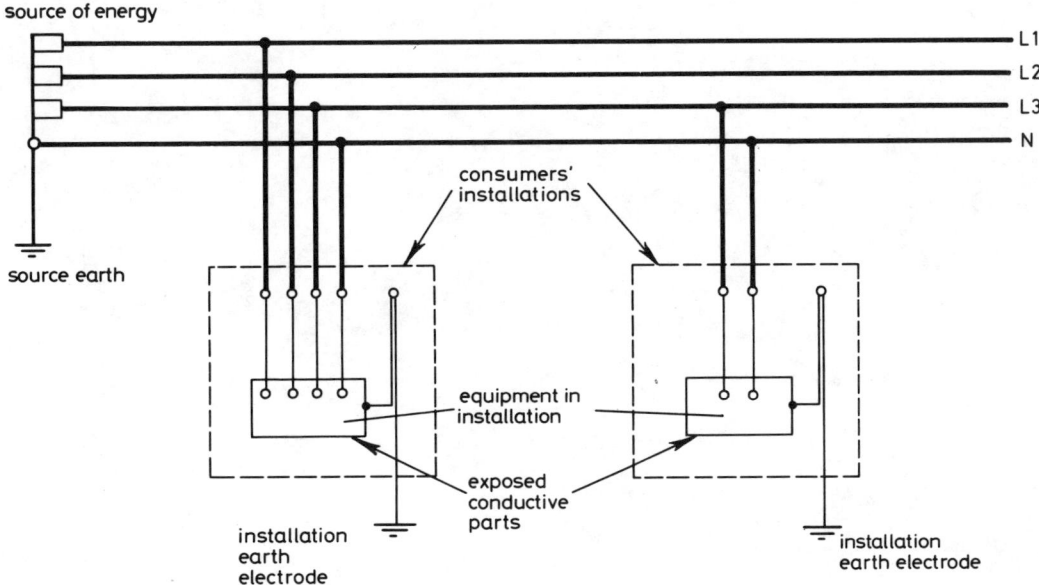

Fig. 6   TT system

All exposed-conductive-parts of an installation are connected to an earth electrode which is electrically independent of the source earth.

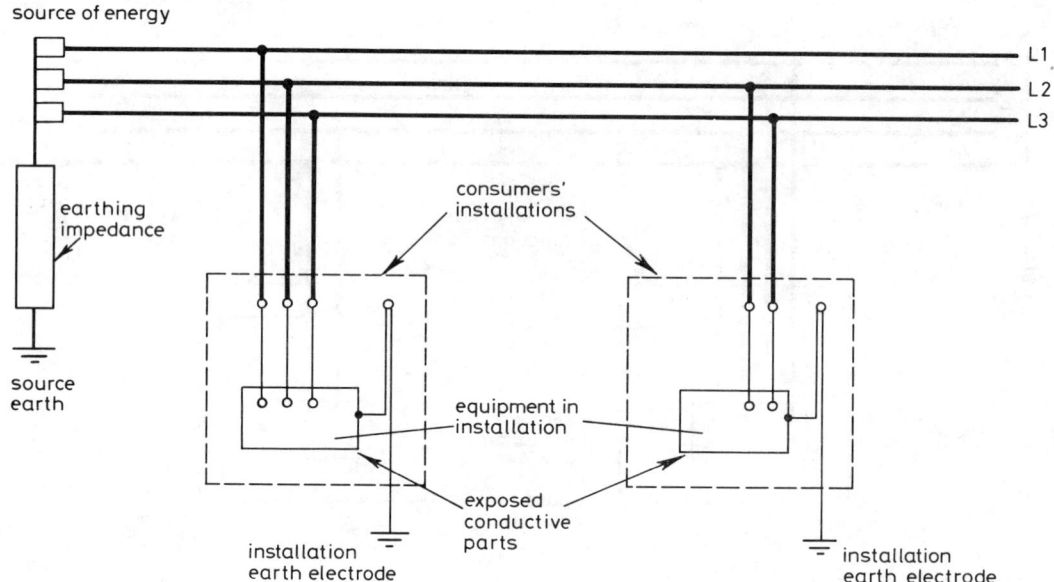

Fig. 7 IT system

All exposed-conductive-parts of an installation are connected to an earth electrode.

The source is either connected to Earth through a deliberately introduced earthing impedance or is isolated from Earth.

PART 3

ASSESSMENT OF GENERAL CHARACTERISTICS

CONTENTS

300-01	General
CHAPTER 31	PURPOSE, SUPPLIES AND STRUCTURE
311	MAXIMUM DEMAND
312	ARRANGEMENT OF LIVE CONDUCTORS AND TYPE OF EARTHING
312-01	General
312-02	Number and type of live conductors
312-03	Type of earthing arrangement
313	NATURE OF SUPPLY
313-01	General
313-02	Supplies for safety services and standby purposes
314	INSTALLATION CIRCUIT ARRANGEMENT
CHAPTER 32	EXTERNAL INFLUENCES
CHAPTER 33	COMPATIBILITY
331-01	General
CHAPTER 34	MAINTAINABILITY
341-01	General

PART 3

ASSESSMENT OF GENERAL CHARACTERISTICS

300-01 General

300-01-01 An assessment shall be made of the following characteristics of the installation in accordance with the Chapters indicated:

(i) the purpose for which the installation is intended to be used, its general structure, and its supplies (Chapter 31)

(ii) the external influences to which it is to be exposed (Chapter 32)

(iii) the compatibility of its equipment (Chapter 33)

(iv) its maintainability (Chapter 34).

These characteristics shall be taken into account in the choice of methods of protection for safety (see Part 4) and the selection and erection of equipment (see Part 5).

CHAPTER 31

PURPOSE, SUPPLIES AND STRUCTURE

311   MAXIMUM DEMAND

311-01-01 The maximum demand of an installation, expressed in amperes, shall be assessed. In determining the maximum demand of an installation or part thereof, diversity may be taken into account.

312   ARRANGEMENT OF LIVE CONDUCTORS AND TYPE OF EARTHING

312-01 General

312-01-01 The characteristics mentioned in Regulations 312-02-01 and 312-03-01 shall be ascertained and appropriate methods of protection for safety selected in compliance with Part 4.

312-02 Number and type of live conductors

312-02-01 The number and type of live conductors (e.g single-phase two-wire a.c., three-phase four-wire a.c.) shall be determined, both for the source of energy and for each circuit to be used within the installation. Where the source of energy is provided by a supplier, the supplier shall be consulted if necessary.

312-03 Type of earthing arrangement

312-03-01 The type of earthing system to be used for the installation shall be determined, due account being taken of the characteristics of the source of energy, and in particular of any facilities for earthing.

Types of system are:

TN-C, TN-S, TN-C-S, TT and IT.

313 NATURE OF SUPPLY

313-01 General

313-01-01 The following characteristics of the supply or supplies, from whatever source, and where appropriate the normal range of those characteristics, shall be determined by calculation, measurement, enquiry or inspection:

  (i) the nominal voltage(s)

  (ii) the nature of the current and frequency

  (iii) the prospective short-circuit current at the origin of the installation

  (iv) the earth fault loop impedance of that part of the system external to the installation (Z_e)

  (v) the suitability for the requirements of the installation, including the maximum demand

  (vi) the type and rating of the overcurrent protective device acting at the origin of the installation.

313-02 Supplies for safety services and standby purposes

313-02-01 Where a supply for safety services or standby purposes is specified, the characteristics of the source or sources of any such supply shall be determined. Such supplies shall have adequate capacity, reliability and rating and appropriate changeover time for the operation specified.

Where the normal source of energy is to be provided by a supplier, the supplier shall be consulted regarding switching arrangements for safety and standby supplies and in particular where the various sources are intended to operate in parallel (or the sources must be prevented from doing so).

Supplies for safety services shall comply with the requirements of Chapter 56.

A source for safety services shall be classified as either automatic or non-automatic. The former may be continuously available (no break) or be available within a stated time delay.

314 INSTALLATION CIRCUIT ARRANGEMENT

314-01-01 Every installation shall be divided into circuits as necessary to:

  (i) avoid danger in the event of a fault, and

  (ii) facilitate safe operation, inspection, testing, and maintenance.

314-01-02 A separate circuit shall be provided for each part of the installation which needs to be separately controlled for compliance with the Regulations or otherwise to prevent danger, so that such circuits remain energised in the event of failure of any other circuit of the installation, and due account shall be taken of the consequences of the operation of any single protective device.

314-01-03 The number of final circuits required, and the number of points supplied by any final circuit, shall be such as to facilitate compliance with the requirements of Chapter 43 for overcurrent protection, Chapter 46 for isolation and switching and Chapter 52 as regards current-carrying capacities of conductors.

314-01-04 Where an installation comprises more than one final circuit, each final circuit shall be connected to a separate way in a distribution board. The wiring of each final circuit shall be electrically separate from that of every other final circuit, so as to prevent the indirect energising of a final circuit intended to be isolated.

CHAPTER 32

EXTERNAL INFLUENCES

At the time of issue of this Edition of the Regulations, the work on requirements for application of the classification of external influences for IEC Publication 364 is insufficiently advanced for adoption as a basis for national regulations. At a later stage, the IEC classification will be considered for adoption as Chapter 32 of these Regulations, together with the proposed IEC requirements for its application.

For the present, Appendix 5 contains some information on the subject and a concise list of external influences which may need to be taken into account.

CHAPTER 33

COMPATIBILITY

331-01 General

331-01-01 An assessment shall be made of any characteristics of equipment likely to have harmful effects upon other electrical equipment or other services, or be harmfully affected by them, or likely to impair the supply.

For an external source of energy the supplier shall be consulted regarding any equipment of the installation having a characteristic likely to have significant influence on the supply.

CHAPTER 34

MAINTAINABILITY

341-01 General

341-01-01 An assessment shall be made of the frequency and quality of maintenance that the installation can reasonably be expected to receive during its intended life. This assessment shall, wherever practicable, include consultation with the person or body who will be responsible for the operation and maintenance of the installation.

Having regard to the frequency and quality of maintenance expected, the requirements of Parts 4 to 7 shall be applied so that during the life of the installation:

(i) any periodic inspection, testing, maintenance and repair likely to be necessary can be readily and safely carried out, and

(ii) any protective measure for safety remains effective, and

(iii) the reliability of equipment is sustained.

PART 4

PROTECTION FOR SAFETY

CONTENTS

400-01	**General**
400-02	**Protective measures for installations and locations of increased shock risk**
CHAPTER 41	**PROTECTION AGAINST ELECTRIC SHOCK**
410-01	**General**
411	PROTECTION AGAINST BOTH DIRECT AND INDIRECT CONTACT
412	PROTECTION AGAINST DIRECT CONTACT
413	PROTECTION AGAINST INDIRECT CONTACT
CHAPTER 42	**PROTECTION AGAINST THERMAL EFFECTS**
421	*(Reserved for future use)*
422	PROTECTION AGAINST FIRE AND HARMFUL THERMAL EFFECTS
423	PROTECTION AGAINST BURNS
CHAPTER 43	**PROTECTION AGAINST OVERCURRENT**
431	GENERAL
432	NATURE OF PROTECTIVE DEVICES
433	PROTECTION AGAINST OVERLOAD CURRENT
434	PROTECTION AGAINST FAULT CURRENT
435	CO-ORDINATION OF OVERLOAD CURRENT AND FAULT CURRENT PROTECTION
436	LIMITATION OF OVERCURRENT BY CHARACTERISTICS OF SUPPLY
CHAPTER 44	*(Reserved for future use)*
CHAPTER 45	**PROTECTION AGAINST UNDERVOLTAGE**
CHAPTER 46	**ISOLATION AND SWITCHING**

460	GENERAL	
461	ISOLATION	
462	SWITCHING OFF FOR MECHANICAL MAINTENANCE	
463	EMERGENCY SWITCHING	
CHAPTER 47	**APPLICATION OF PROTECTIVE MEASURES FOR SAFETY**	
470	GENERAL	
471	PROTECTION AGAINST ELECTRIC SHOCK	
472	*(Reserved for future use)*	
473	PROTECTION AGAINST OVERCURRENT	
474	*(Reserved for future use)*	
475	*(Reserved for future use)*	
476	ISOLATION AND SWITCHING	

CHAPTER 41

PROTECTION AGAINST ELECTRIC SHOCK

CONTENTS

410-01	**General**
411	PROTECTION AGAINST BOTH DIRECT AND INDIRECT CONTACT
411-01	**General**
411-02	**Protection by SELV**
411-02-01 to 411-02-04	Sources for SELV
411-02-05 to 411-02-11	Arrangement of circuits for SELV
411-03	**Other Extra-Low voltage systems including Functional extra-low voltage.**
411-04	**Protection by limitation of discharge of energy**
412	PROTECTION AGAINST DIRECT CONTACT
412-01	**General**
412-02	**Protection by insulation of live parts**
412-03	**Protection by barriers or enclosures**
412-04	**Protection by obstacles**
412-05	**Protection by placing out of reach**
412-06	**Supplementary protection by residual current devices**
413	PROTECTION AGAINST INDIRECT CONTACT
413-01	**General**
413-02	**Protection by earthed equipotential bonding and automatic disconnection of supply**
413-02-01 to 413-02-05	General
413-02-06 to 413-02-17	TN System

413-02-18 to **413-02-20**	TT System
413-02-21 to **413-02-26**	IT System
413-02-27 and **413-02-28**	Supplementary equipotential bonding
413-03	**Protection by use of Class II equipment or by equivalent insulation**
413-04	**Protection by non-conducting location**
413-05	**Protection by earth-free local equipotential bonding**
413-06	**Protection by electrical separation**

PART 4

PROTECTION FOR SAFETY

400-01 General

400-01-01 Every installation, either as a whole or in its several parts, shall comply with the requirements of this Part by the application in accordance with Chapter 47 of the protective measures described in Chapters 41 to 46.

The order in which the protective measures are listed shall not be taken to imply the relative importance of different measures.

400-02 Protective measures for installations and locations of increased shock risk

400-02-01 For any installation or part of an installation where the risk of electric shock is increased by a reduction in body resistance or by contact with earth potential, the relevant requirements of Part 6 of these Regulations shall be applied in addition to the requirements contained in other parts of the Regulations.

CHAPTER 41

PROTECTION AGAINST ELECTRIC SHOCK

410-01 General

410-01-01 Protection against electric shock shall be provided by the application, in accordance with Section 471, of:

(i) an appropriate measure specified in Section 411, for protection against both direct contact and indirect contact, or

(ii) a combination of appropriate measures specified in Section 412 for protection against direct contact and Section 413 for protection against indirect contact.

411 PROTECTION AGAINST BOTH DIRECT AND INDIRECT CONTACT

411-01 General

411-01-01 For protection against both direct contact and indirect contact one of the following basic protective measures shall be used:

(i) protection by SELV according to Regulations 411-02 and 471-02

(ii) protection by limitation of discharge of energy according to Regulations 411-04 and 471-03.

411-01-02 Functional extra-low voltage alone shall not be used as a protective measure. (See Regulation 471-14.)

411-02 Protection by SELV

Sources for SELV

411-02-01 For protection by SELV, compliance with all the following requirements shall be provided:

(i) the nominal voltage of the circuit concerned shall not exceed extra-low voltage

(ii) the supply shall be from one of the safety sources listed in Regulation 411-02-02

(iii) the conditions of Regulations 411-02-03 to 411-02-11 shall be fulfilled.

411-02-02 The source for SELV shall be one of the following:

(i) a safety isolating transformer complying with BS 3535 in which there shall be no connection between the output winding and the body or the protective earthing circuit, if any

(ii) a source of current such as a motor generator with windings providing electrical separation equivalent to that of the safety isolating transformer specified in (i) above

(iii) an electrochemical source (e.g. a battery) or another source independent of a higher voltage circuit (e.g. an engine-driven generator)

(iv) certain electronic devices complying with appropriate standards where measures have been taken so that even in the case of an internal fault the voltage at the output terminals cannot exceed the value specified by Regulation 411-02-01 (i). A higher voltage at the output terminals is, however, permitted if, in the case of direct or indirect contact, the voltage at the output terminals is immediately reduced to the value specified by Regulation 411-02-01(i). Where such a higher voltage exists, compliance with this item (iv) is deemed to be met if the voltage at the output terminals is within the limit specified by Regulation 411-02-01(i) when measured with a voltmeter having an internal resistance of at least 3000 ohms.

411-02-03 A system supplied from a higher voltage system by other equipment which does not provide the necessary electrical separation, such as an autotransformer, potentiometer, semiconductor device etc., shall not be deemed to be a SELV system.

411-02-04 A mobile source for SELV shall be selected and erected in accordance with the requirements for protection by the use of Class II equipment or by equivalent insulation (see Regulations 413-03 and 471-09).

Arrangement of circuits for SELV

411-02-05 A live part of a SELV system shall:

(i) be electrically separated from that of any other higher voltage system. Except for cables the electrical separation shall be not less than that between the input and output of a safety isolating transformer

(ii) not be connected to Earth or to a live part or a protective conductor forming part of another system.

411-02-06 Circuit conductors of each SELV system shall be physically separated from those of any other system or, where compliance with this requirement is impracticable, installed in accordance with one of the following arrangements:

(i) SELV circuit conductors shall be insulated in accordance with the requirements of the Regulations for the highest voltage present

(ii) SELV circuit cables shall be non-metallic sheathed cables complying with Chapter 52

(iii) conductors of systems at voltages higher than SELV shall be separated from those at SELV by an earthed metallic screen or an earthed metallic sheath

(iv) where circuits at different voltages are contained in a multicore cable or other grouping of conductors, the conductors of SELV circuits shall be insulated, individually or collectively, for the highest voltage present in the cable or grouping.

In the arrangements (ii) and (iii) above, basic insulation of any conductor need be sufficient only for the voltage of the circuit of which it is a part.

411-02-07 No exposed-conductive-part of a SELV system shall be connected to any of the following:

(i) Earth

(ii) an exposed-conductive-part of another system

(iii) a protective conductor of any system

(iv) an extraneous-conductive-part, except that where electrical equipment is inherently required to be connected to an extraneous-conductive-part, measures shall be incorporated so that the parts cannot attain a voltage exceeding the appropriate value specified by Regulation 411-02-01(i).

411-02-08 If an exposed-conductive-part of an extra-low voltage system is liable to come into contact, either fortuitously or intentionally, with an exposed-conductive-part of any other system, the system shall not be deemed to be SELV and other protective measures shall be applied in accordance with Regulation 411-03.

411-02-09 If the nominal voltage of a SELV system exceeds 25 V r.m.s. a.c. or 60 V ripple-free d.c., protection against direct contact shall be provided by one or more of the following:

(i) a barrier or an enclosure (Regulation 412-03) affording at least the degree of protection IP 2X (see BS 5490)

(ii) insulation (Regulation 412-02) capable of withstanding a test voltage of 500 V r.m.s. a.c. for 1 minute.

If the nominal voltage does not exceed 25 V r.m.s. a.c. or 60 V ripple-free d.c., protection as described in (i) and (ii) above is unnecessary.

For the purposes of this Regulation "ripple-free" means, for sinusoidal ripple voltage, a ripple content not exceeding 10% r.m.s.; the maximum peak value shall not exceed 140 V for a nominal 120 V ripple-free d.c. system and 70 V for a nominal 60 V ripple-free d.c. system.

411-02-10 A socket-outlet in a SELV system shall require the use of a plug which is not dimensionally compatible with those used for any other system in use in the same premises and shall not have a protective conductor contact.

411-02-11 A luminaire supporting coupler having a protective conductor contact shall not be installed in a SELV system.

411-03 Other extra-low voltage systems including functional extra-low Voltage

411-03-01 Where, for functional reasons, Extra-low voltage is used but not all of the requirements relating to SELV are fulfilled, other measures shall be applied to provide protection against direct and indirect contact, as specified in Regulation 471-14.

411-04 Protection by limitation of discharge of energy

411-04-01 For equipment complying with the appropriate British Standard, protection against both direct and indirect contact shall be deemed to be provided when the equipment incorporates means of limiting the current which can pass through the body of a person or livestock to a value lower than that likely to cause danger.

A circuit relying on this protective measure shall be separated from any other circuit in a manner equivalent to that specified in Regulations 411-02-05 and 411-02-06 for a SELV circuit.

412 PROTECTION AGAINST DIRECT CONTACT

412-01 General

412-01-01 For protection against direct contact one or more of the following basic protective measures shall be used in accordance with the relevant requirements of this Section and the application rules specified in Section 471, including those indicated below:

(i) protection by insulation of live parts (Regulation 412-02 and Regulation 471-04)

(ii) protection by a barrier or an enclosure (Regulation 412-03 and Regulation 471-05)

(iii) protection by obstacles (Regulation 412-04 and Regulation 471-06)

(iv) protection by placing out of reach (Regulation 412-05 and Regulation 471-07).

Where supplementary protection by a residual current device is also provided, the requirements of Regulation 412-06 shall apply.

412-02 Protection by insulation of live parts (see also Regulation 471-04)

412-02-01 Live parts shall be completely covered with insulation which can only be removed by destruction and which is capable of durably withstanding the electrical, mechanical, thermal and chemical stresses to which it may be subjected in service.

Where insulation is applied during the erection of the installation, the quality of the insulation shall be verified by tests equivalent to those specified in the British Standard for similar type-tested equipment.

412-03   Protection by barriers or enclosures (see also Regulation 471-05)

412-03-01   Live parts shall be inside enclosures or behind barriers providing at least the degree of protection IP 2X (see BS 5490) except that, where an opening larger than that permitted for IP 2X is necessary to allow the replacement of parts or to avoid interference with the proper functioning of electrical equipment both of the following requirements apply:

(i) suitable precautions shall be taken to prevent persons or livestock from touching a live part unintentionally

(ii) it shall be established, as far as practicable, that a person will be aware that a live part can be touched through the opening and should not be touched (see Regulation 471-05-02).

412-03-02   The horizontal top surface of a barrier or enclosure which is readily accessible shall provide a degree of protection of at least IP 4X (see BS 5490).

412-03-03   Every barrier and enclosure shall be firmly secured in place and have sufficient stability and durability to maintain the required degree of protection and appropriate separation from any live part in the known conditions of normal service.

412-03-04   Where it is necessary to remove a barrier or to open an enclosure or to remove a part of an enclosure, one or more of the following requirements shall be satisfied:

(i) the removal or opening shall be possible only by use of a key or tool

(ii) the removal or opening shall be possible only after disconnection of the supply to the live part against which the barrier or enclosure affords protection, restoration of the supply being possible only after replacement or reclosure of the barrier or enclosure

(iii) an intermediate barrier shall be provided to prevent contact with a live part, such a barrier affording a degree of protection of at least IP 2X (see BS 5490) and removable only by the use of a tool.

This Regulation does not apply to a ceiling rose complying with BS 67 or to a pull cord switch complying with BS 3676, nor to a bayonet lampholder complying with BS 5042 nor to an Edison screw lampholder complying with BS 6776.

412-04   Protection by obstacles (see also Regulation 471-06)

412-04-01   An obstacle shall prevent, as appropriate, either of the following:

(i) unintentional bodily approach to a live part

(ii) unintentional contact with a live part when operating energised equipment.

412-04-02   An obstacle shall be so secured as to prevent unintentional removal, but may be removable without using a key or tool.

412-05   Protection by placing out of reach (see also Regulation 471-07)

412-05-01   A bare or insulated overhead line for distribution between buildings and structures shall be installed to the standard required by the Electricity Supply Regulations 1988, (as amended).

412-05-02   A bare live part other than an overhead line shall not be within arm's reach or 2.5 m of any of the following:

(i) an exposed-conductive-part
(ii) an extraneous-conductive-part
(iii) a bare live part of any other circuit.

412-05-03   If access to live equipment from a normally occupied position is restricted in the horizontal plane by an obstacle (e.g. handrail, mesh, screen) affording a degree of protection less than IP 2X, (see BS 5490) the extent of arm's reach shall be measured from that obstacle.

412-05-04   In each place where any bulky or long conducting object is normally handled, the distances required by Regulations 412-05-02 and 412-05-03 shall be increased accordingly.

412-06 Supplementary protection by residual current devices

412-06-01 A residual current device shall not be used as the sole means of protection against direct contact.

412-06-02 The use of a residual current device is recognised as reducing the risk of electric shock where the following conditions are complied with:

(i) one of the protective measures specified in Items (i) to (iv) of Regulation 412-01-01 shall be applied, and

(ii) the residual current device shall have a rated residual operating current not exceeding 30 mA and an operating time not exceeding 40 ms at a residual current of 150 mA, as provided by BS 4293.

413 PROTECTION AGAINST INDIRECT CONTACT

413-01 General

413-01-01 For protection against indirect contact, one of the following basic measures shall be used in accordance with the relevant requirements of this Section and the application rules specified in Section 471, including those indicated below:

(i) protection by earthed equipotential bonding and automatic disconnection of supply (Regulation 413-02 and Regulation 471-08)

(ii) protection by Class II equipment or by equivalent insulation (Regulation 413-03 and Regulation 471-09)

(iii) protection by non-conducting location (Regulation 413-04 and Regulation 471-10)

(iv) protection by earth-free local equipotential bonding (Regulation 413-05 and Regulation 471-11)

(v) protection by electrical separation (Regulation 413-06 and Regulation 471-12).

413-02 Protection by earthed equipotential bonding and automatic disconnection of supply
(see also Regulation 471-08)

General

413-02-01 This measure shall be applied in accordance with the requirements for the type of system earthing in use:

(i) TN systems: Regulations 413-02-06 to 413-02-17
(ii) TT systems: Regulations 413-02-18 to 413-02-20
(iii) IT systems: Regulations 413-02-21 to 413-02-26

413-02-02 In each installation main equipotential bonding conductors complying with Section 547 shall connect to the main earthing terminal for that installation extraneous-conductive-parts including the following:

(i) water service pipes
ii) gas installation pipes
(iii) other service pipes and ducting
(iv) central heating and air conditioning systems
(v) exposed metallic structural parts of the building
(vi) the lightning protective system.

Where an installation serves more than one building the above requirement shall be applied to each building.

To comply with the Regulations it is also necessary to apply equipotential bonding to any metallic sheath of a telecommunication cable. However the consent of the owner or operator of the cable shall be obtained.

413-02-03 Simultaneously accessible exposed-conductive-parts shall be connected to the same earthing system individually, in groups or collectively.

413-02-04 The characteristics of each protective device for automatic disconnection, the earthing arrangements for the installation and the relevant impedance of the circuit concerned shall be co-ordinated so that during an earth fault the voltages between simultaneously accessible exposed and extraneous-conductive-parts occurring anywhere in the installation shall be of such magnitude and duration as not to cause danger.

Conventional means of compliance with this Regulation are given in Regulations 413-02-06 to 413-02-26 according to the type of system earthing, but other equally effective means shall not be excluded.

413-02-05 Account shall be taken of the effect on the resistance of every circuit conductor of its temperature rise during the clearance of the fault.

TN system

413-02-06 Each exposed-conductive-part of the installation shall be connected by a protective conductor to the main earthing terminal of the installation and that terminal shall be connected to the earthed point of the supply source in accordance with Regulations 542-01-02, 542-01-03 and 542-01-05, as appropriate.

413-02-07 One or more of the following types of protective device shall be used:

(i) an overcurrent protective device
(ii) a residual current device.

Where a residual current device is used in a TN-C-S system, a PEN conductor shall not be used on the load side. Connection of the protective conductor to the PEN conductor shall be made on the source side of the residual current device.

413-02-08 Regulation 413-02-04 is considered to be satisfied if the characteristic of each protective device and the earth fault loop impedance of each circuit protected by it are such that automatic disconnection of the supply will occur within a specified time when a fault of negligible impedance occurs between a phase conductor and a protective conductor or an exposed-conductive-part anywhere in the installation. This requirement is met where the following condition is fulfilled:

$$Z_s \leq \frac{U_o}{I_a}$$

Where:

Z_s is the earth fault loop impedance.

I_a is the current causing the automatic operation of the disconnecting protective device within the time stated in Table 41A as a function of the nominal voltage U_o or, under the conditions stated in Regulations 413-02-12 and 413-02-13, within a time not exceeding 5 s.

U_o is the nominal voltage to Earth.

TABLE 41A

Maximum disconnection times for TN Systems

(see Regulation 413-02-09)

U_o (volts)	t (seconds)
120	0.8
220 to 277	0.4
400	0.2
greater than 400	0.1

413-02-09 The maximum disconnection times of Table 41A shall apply to a circuit supplying socket-outlets and to other final circuits which supply portable equipment intended for manual movement during use, or hand-held Class I equipment.

This requirement does not apply to a final circuit supplying an item of stationary equipment connected by means of a plug and socket-outlet where precautions are taken to prevent the use of the socket-outlet for supplying hand-held equipment, nor to the reduced low-voltage circuits described in Regulation 471-15.

413-02-10 Where a fuse is used to satisfy the requirements of Regulation 413-02-09, maximum values of earth fault loop impedance (Z_s) corresponding to a disconnection time of 0.4 s are stated in Table 41B1 for a nominal voltage to Earth (U_o) of 240 V. For types and rated currents of general purpose (gG) fuses other than those mentioned in Table 41B1, and for motor circuit fuses (gM), reference should be made to the appropriate British Standard, to determine the value of I_a for compliance with Regulation 413-02-08.

TABLE 41B1

Maximum earth fault loop impedance (Z_s) for fuses, for 0.4 s disconnection time with U_O 240 V (see Regulation 413-02-10)

(a) General purpose (gG) fuses to BS 88 Parts 2 and 6

Rating (amperes)	6	10	16	20	25	32	40	50
Z_s (ohms)	8.89	5.33	2.82	1.85	1.50	1.09	0.86	0.63

(b) Fuses to BS 1361

Rating (amperes)	5	15	20	30	45
Z_s (ohms)	10.9	3.43	1.78	1.20	0.60

(c) Fuses to BS 3036

Rating (amperes)	5	15	20	30	45
Z_s (ohms)	10.0	2.67	1.85	1.14	0.62

(d) Fuses to BS 1362

Rating (amperes)	13
Z_s (ohms)	2.53

413-02-11 Where a circuit-breaker is used to satisfy the requirements of Regulation 413-02-09, the maximum value of earth fault loop impedance (Z_s) shall be determined by the formula of Regulation 413-02-08. Alternatively, for a nominal voltage to earth of 240 V and a disconnection time of 0.4 s, the values specified in Table 41B2 for the types and ratings of circuit-breaker listed may be used instead of calculation.

TABLE 41B2

Maximum earth fault loop impedance (Z_s) for miniature circuit-breakers, for disconnection times of both 0.4 s with U_O 240 V (see Regulation 413-02-11) and 5 s (see Regulations 413-02-12 and 413-02-14)

(e) Type 1 miniature circuit-breakers to BS 3871

Rating (amperes)	5	6	10	15	16	20	30	32	40	45	50	63	I_n
Z_s (ohms)	12	10	6	4	3.75	3	2	1.88	1.5	1.33	1.2	0.95	240/(4I_n)

(f) Type 2 miniature circuit-breakers to BS 3871

Rating (amperes)	5	6	10	15	16	20	30	32	40	45	50	63	I_n
Z_s (ohms)	6.86	5.71	3.43	2.29	2.14	1.71	1.14	1.07	0.86	0.76	0.69	0.54	240/(7I_n)

(g) Type B miniature circuit-breakers to BS 3871

Rating (amperes)	6	10	16	20	32	40	45	50	63	I_n
Z_s (ohms)	8.0	4.80	3.0	2.40	1.50	1.20	1.07	0.96	0.76	240/(5I_n)

(h) Type 3 and Type C miniature circuit-breakers to BS 3871

Rating (amperes)	5	6	10	15	16	20	30	32	40	45	50	63	I_n
Z_s (ohms)	4.80	4.00	2.40	1.60	1.50	1.20	0.80	0.75	0.60	0.53	0.48	0.38	240/(10I_n)

413-02-12 Irrespective of the value of U_o, for a final circuit which supplies a socket-outlet or portable equipment intended for manual movement during use, or hand-held Class I equipment, it shall be permissible to increase the disconnection time to a value not exceeding 5 s for the types and ratings of the overcurrent protective devices and associated maximum impedances of the circuit protective conductors shown in Table 41C. The impedance of the protective conductor shall be referred to the point of connection to the main equipotential bonding. Where additional equipotential bonding is installed in accordance with Regulation 413-02-13 (ii) then the impedance of the circuit protective conductor specified in this paragraph applies to that portion of the circuit protective conductor between the point of additional bonding and the socket-outlet, or portable equipment.

TABLE 41C

Maximum impedance of circuit protective conductor related to the final circuit protective device
(see Regulation 413-02-12)

(a) General purpose (gG) fuses to BS 88 Parts 2 and 6

Rating (amperes)	6	10	16	20	25	32	40	50
Impedance (ohms)	2.94	1.61	0.91	0.63	0.50	0.40	0.29	0.23

(b) Fuses to BS 1361

Rating (amperes)	5	15	20	30	45
Impedance (ohms)	3.57	1.09	0.61	0.40	0.21

(c) Fuses to BS 3036

Rating (amperes)	5	15	20	30	45
Impedance (ohms)	3.85	1.16	0.83	0.58	0.35

(d) Fuses to BS 1362

Rating (amperes)	13
Impedance (ohms)	0.83

(e) Type 1 miniature circuit-breakers to BS 3871

Rating (amperes)	5	6	10	15	16	20	30	32	40	45	50	63	I_n
Impedance (ohms)	2.50		1.25		0.78		0.42		0.31		0.25		$12.5/I_n$
		2.08		0.83		0.63		0.39		0.28		0.2	

(f) Type 2 miniature circuit-breakers to BS 3871

Rating (amperes)	5	6	10	15	16	20	30	32	40	45	50	63	I_n
Impedance (ohms)		1.19		0.48		0.36		0.22		0.16		0.11	
	1.43		0.71		0.45		0.24		0.18		0.14		$50/(7I_n)$

(g) Type B miniature circuit-breakers to BS 3871

Rating (amperes)	6	10	16	20	32	40	45	50	63	I_n
Impedance (ohms)	1.67		0.63		0.31		0.22		0.16	
		1.00		0.50		0.25		0.20		$10/I_n$

(h) Type 3 and Type C miniature circuit-breakers to BS 3871

Rating (amperes)	5	6	10	15	16	20	30	32	40	45	50	63	I_n
Impedance (ohms)	1.00	0.83	0.50	0.33	0.31	0.25	0.17	0.16	0.13	0.11	0.10	0.08	$5/I_n$

413-02-13 For a distribution circuit a disconnection time not exceeding 5 s is permitted.

For a final circuit supplying only stationary equipment and for a final circuit for which the requirement of Regulation 413-02-09 does not apply a disconnection time not exceeding 5 s is permitted. Where the disconnecting time for such a final circuit exceeds that required by Table 41A and another final circuit requiring a disconnection time according to Table 41A is connected to the same distribution board or distribution circuit, one of the following conditions shall be fulfilled:

(i) the impedance of the protective conductor between the distribution board and the point at which the protective conductor is connected to the main equipotential bonding shall not exceed the value given in Table 41C for the appropriate protective device in the final circuit, or, for protective devices not included in Table 41C, $50Z_s/U_o$ ohms (where Z_s is the earth fault loop impedance corresponding to a disconnection time of 5 s)

(ii) there shall be equipotential bonding at the distribution board, which involves the same types of extraneous-conductive-parts as the main equipotential bonding according to Regulation 413-02-02 and is sized in accordance with Regulation 547-02-01.

413-02-14 Where a circuit-breaker is used to satisfy the requirements of Regulation 413-02-13, the maximum value of earth fault loop impedance (Z_s) shall be determined by the formula of Regulation 413-02-08. Alternatively the values specified in Table 41B2 may be used instead of calculation for a nominal voltage to Earth (U_o) of 240 V for the types and ratings of circuit-breaker listed therein.

Where a fuse is used to satisfy the requirements of Regulation 413-02-13, maximum values of earth fault loop impedance (Z_s) corresponding to a disconnection time of 5 s are stated in Table 41D for a nominal voltage to Earth (U_o) of 240 V.

For types and rated currents of general purpose (gG) fuses other than those mentioned in Table 41D and for motor circuit fuses (gM) reference should be made to the appropriate British Standard to determine the value of I_a for compliance with Regulation 413-02-08.

TABLE 41D

Maximum earth fault loop impedance (Z_s) for 5 s disconnection time with U_o 240 V
(see Regulations 413-02-13 and 413-02-14)

(a) General purpose (gG) fuses to BS 88 Parts 2 and 6

Rating (amperes)	6	10	16	20	25	32	40	50
Z_s (ohms)	14.1	7.74	4.36	3.04	2.40	1.92	1.41	1.09

Rating (amperes)	63	80	100	125	160	200
Z_s (ohms)	0.86	0.60	0.44	0.35	0.27	0.20

(b) Fuses to BS 1361

Rating (amperes)	5	15	20	30	45	60	80	100
Z_s (ohms)	17.1	5.22	2.93	1.92	1.00	0.73	0.52	0.38

(c) Fuses to BS 3036

Rating (amperes)	5	15	20	30	45	60	100
Z_s (ohms)	18.5	5.58	4.00	2.76	1.66	1.17	0.56

(d) Fuse to BS 1362

Rating (amperes)	13
Z_s (ohms)	4

413-02-15 Where the conditions of Regulations 413-02-08 to 413-02-14, for automatic disconnection, cannot be fulfilled by using overcurrent protective devices, then either:

(i) local supplementary equipotential bonding shall be applied in accordance with Regulations 413-02-27 and 413-02-28, but the use of such bonding does not obviate the need to disconnect the supply for reasons other than protection against electric shock, such as thermal effects, or

(ii) protection shall be provided by means of a residual current device.

413-02-16 If protection is provided by a residual current device the following condition shall be fulfilled:

$$Z_s I_n \leq 50 \text{ V}$$

Where:

Z_s is the earth fault loop impedance in ohms.

I_n is the rated residual operating current of the protective device in amperes.

413-02-17 If a residual current device is used for automatic disconnection for a circuit which extends beyond the earthed equipotential zone, exposed-conductive-parts need not be connected to the TN system protective conductors provided that they are connected to an earth electrode affording a resistance appropriate to the operating current of the residual current device. The circuit thus protected is to be treated as a TT system and Regulations 413-02-18 to 413-02-20 apply.

TT system

413-02-18 Every exposed-conductive-part which is to be protected by a single protective device shall be connected, via the main earthing terminal, to a common earth electrode. However if several protective devices are in series, the exposed-conductive-parts may be connected to separate earth electrodes corresponding to each protective device.

413-02-19 One or more of the following types of protective device shall be used, the former being preferred:

(i) a residual current device

(ii) an overcurrent protective device.

413-02-20 The following condition shall be fulfilled for each circuit:

$$R_a I_a \leq 50 \text{ V}$$

Where:

R_a is the sum of the resistances of the earth electrode and the protective conductor(s) connecting it to the exposed-conductive-part.

I_a is the current causing the automatic operation of the protective device within 5 s;

When the protective device is a residual current device, I_a is the rated residual operating current I_n.

IT system

413-02-21 No live conductor shall be directly connected with Earth. Where a connection with Earth is required it shall be through a high impedance such that in the event of a single fault to an exposed-conductive-part or to Earth the fault current is so low that it will not give rise to the risk of an electric shock. If a high impedance connection with Earth is made at the neutral point or at an artificial star point or to one of the live conductors, the relevant earth fault loop impedance shall satisfy that condition.

Where, to reduce overvoltage or to damp voltage oscillation, it is necessary to provide earthing through an impedance, its characteristics shall be appropriate to the requirements of the installation.

413-02-22 Disconnection in the event of a single fault to an exposed-conductive-part or to Earth is not imperative if the conditions of Regulation 413-02-21 are satisfied, but precautions shall be taken to guard against the risk of electric shock in the event of two faults existing simultaneously. Either or both of the following devices shall be used:

(i) an overcurrent protective device
(ii) a residual current protective device.

413-02-23 Every exposed-conductive-part shall be earthed such that the following condition shall be fulfilled for each circuit:

$$R_b \, I_d \leq 50 \text{ V}$$

Where:

R_b is the resistance of the earth electrode for exposed-conductive-parts.

I_d is the fault current of the first fault of negligible impedance between a phase conductor and an exposed-conductive-part. The value of I_d takes account of leakage currents and the total earthing impedance of the electrical installation.

413-02-24 An insulation monitoring device shall be provided so as to indicate the occurrence of a first fault from a live part to an exposed-conductive-part or to Earth.

413-02-25 After the occurrence of a first fault, conditions for disconnection of supply in the event of a second fault depend on whether all exposed-conductive-parts are connected together by a protective conductor (collectively earthed) or are earthed in groups or individually:

(i) where exposed-conductive-parts are earthed collectively, the conditions for a TN system shall apply, subject to Regulation 413-02-26

(ii) where exposed-conductive-parts are earthed in groups or individually, conditions for protection shall comply with Regulation 413-02-20 as for TT systems.

Where in an IT system protection against indirect contact is provided by a residual current device, each final circuit shall be separately protected.

413-02-26 The following conditions shall be fulfilled where the neutral is not distributed, (three-phase three-wire distribution):

$$Z_s \leq \frac{0.866 \, U_o}{I_a}$$

or, where the neutral is distributed, (three-phase four-wire distribution and single-phase distribution):

$$Z^1_s \leq \frac{0.5 \, U_o}{I_a}$$

Where:

Z_s is the impedance of the earth fault loop comprising the phase conductor and the protective conductor of the circuit.

Z^1_s is the impedance of the earth fault loop comprising the neutral conductor and the protective conductor of the circuit.

I_a is the current which disconnects the circuit within the time t specified in Table 41E when applicable (see Regulation 413-02-09), or within 5 s for all other circuits when this time is allowed (see Regulation 413-02-13).

TABLE 41E

Maximum disconnection time in IT systems (Second fault)

Installation nominal voltage U_o (Volts)	Maximum Disconnection time t(seconds)	
	Neutral not distributed	Neutral distributed
120	0.8	5.0
220 to 277	0.4	0.8
400	0.2	0.4
580	0.1	0.2

Supplementary equipotential bonding

413-02-27 Where supplementary equipotential bonding is necessary for compliance with Regulation 413-02-15 or Part 6, it shall connect together the exposed-conductive-parts of equipment in the circuits concerned including the earthing contacts of socket-outlets and extraneous-conductive-parts in accordance with Regulation 547-03.

413-02-28 The resistance (R) of the supplementary bonding conductor between simultaneously accessible exposed-conductive-parts and extraneous-conductive-parts shall fulfil the following condition:

$$R \leq \frac{50}{I_a}$$

Where:

I_a is the operating current of the protective device:

(i) for a residual current device, I_n

(ii) for an overcurrent device, it is the minimum current which disconnects the circuit within 5 s.

413-03 **Protection by use of Class II equipment or by equivalent insulation** (see also Regulation 471-09)

413-03-01 Protection shall be provided by one or more of the following:

(i) electrical equipment of the following types, type-tested and marked to the relevant standards:

(a) electrical equipment having double or reinforced insulation (Class II equipment)

(b) low-voltage switchgear and controlgear assemblies having total insulation (see BS 5486).

(ii) supplementary insulation applied to electrical equipment having basic insulation only, as a process in the erection of an electrical installation, providing a degree of safety equivalent to that of electrical equipment according to item (i) above and complying with Regulations 413-03-03 to 413-03-09

(iii) reinforced insulation applied to uninsulated live parts, as a process in the erection of an electrical installation, providing a degree of safety equivalent to electrical equipment according to item (i) above and complying with Regulations 413-03-03 to 413-03-09, such insulation being recognised only where constructional features prevent the application of double insulation.

413-03-02 The installation of equipment described in item (i) of Regulation 413-03-01 shall be effected in such a way as not to impair the protection afforded in compliance with the equipment specification. Class II equipment shall be so installed that basic insulation is not the only protection between live parts of the installation and exposed metalwork of that equipment.

413-03-03 The enclosure provided for this measure shall not adversely affect the operation of the equipment protected.

413-03-04 The electrical equipment being ready for operation, all conductive parts separated from live parts only by basic insulation shall be contained in an insulating enclosure affording at least the degree of protection IP 2X (see BS 5490).

413-03-05 The insulating enclosure shall be capable of resisting mechanical, electrical and thermal stresses likely to be encountered.

A coating of paint, varnish or similar product is generally not considered to comply with these requirements. This requirement does not exclude, however, the use of a type-tested enclosure provided with such a coating if the relevant product standard admits its use and if the insulating coating has been tested according to the requirements of the relevant product standard.

413-03-06 If the insulating enclosure has not previously been tested, a suitable test shall be carried out (see Regulation 713-05-02).

413-03-07 The insulating enclosure shall not be pierced by conductive parts, other than circuit conductors, likely to transmit a potential. The insulating enclosure shall not contain any screws of insulating material, the replacement of which by metallic screws could impair the insulation provided by the enclosure.

Where the insulating enclosure has to be pierced by conductive parts (e.g. for operating handles of built-in equipment, and for screws) protection against indirect contact shall not be impaired.

413-03-08 Where a lid or door in an insulating enclosure can be opened without the use of a tool or key, every conductive part which is accessible if the lid or door is open shall be behind an insulating barrier which prevents a person from coming into contact with those parts; this insulating barrier shall provide a degree of protection of at least IP 2X (see BS 5490) and be removable only by use of a tool.

413-03-09 No conductive part enclosed in an insulating enclosure shall be connected to a protective conductor.

Where provision is made within the enclosure for a protective conductor which necessarily runs through the enclosure in order to serve another item of electrical equipment whose supply circuit also runs through the enclosure, any such protective conductor and its terminals and joints shall be insulated as though they were live parts and its terminals shall be appropriately marked.

413-04 Protection by non-conducting location

413-04-01 This method of protection is not recognised for general application (see Regulation 471-10).

413-04-02 Exposed-conductive-parts which might attain different potentials through failure of the basic insulation of live parts shall be arranged so that a person will not come into simultaneous contact with:

(i)   two exposed-conductive-parts, or

(ii) an exposed-conductive-part and any extraneous-conductive-part.

413-04-03 In a non-conducting location there shall be no protective conductors, and any socket-outlet shall not incorporate an earthing contact. A luminaire supporting coupler having a protective conductor contact shall not be installed.

413-04-04 The resistance of an insulating floor or wall at every point of measurement under the conditions specified in Regulation 713-08 shall be not less than:

(i)   50 kΩ where the voltage to earth does not exceed 500 V

(ii) 100 kΩ where the voltage to earth exceeds 500 V but does not exceed low voltage.

If at any point the resistance is less than the specified value, the floors and walls are deemed to be extraneous-conductive-parts for the purposes of protection against electric shock.

413-04-05 Permanent arrangement shall be made which shall afford protection where the use of mobile or portable equipment is envisaged.

413-04-06 Precautions shall be taken so that a potential on extraneous-conductive-parts in the location cannot be transmitted outside that location.

413-04-07 Regulation 413-04-02 shall be deemed to be fulfilled if the location has an insulating floor and insulating walls and one or more of the following arrangements applies:

(i) the distance between any separated exposed-conductive-parts and between exposed-conductive-parts and extraneous-conductive-parts is not less than 2 m or, for parts out of the zone of arm's reach, not less than 1.25 m

(ii) interposition of effective obstacles between exposed-conductive-parts and extraneous-conductive-parts. Such obstacles are sufficiently effective if they extend the distances to be surmounted to the values stated in item (i) above. They shall not be connected to Earth or to exposed-conductive-parts; as far as possible they shall be of insulating material

(iii) insulation or insulating arrangement of extraneous-conductive-parts. The insulation shall be of adequate electrical and mechanical strength.

413-05 Protection by earth-free local equipotential bonding (see also Regulation 471-11)

413-05-01 This method shall be used only in special circumstances.

413-05-02 An equipotential bonding conductor shall connect together every simultaneously accessible exposed-conductive-part and extraneous-conductive-part.

413-05-03 The local equipotential bonding conductors shall not be in electrical contact with Earth.

413-05-04 Precautions shall be taken so that persons entering or leaving the equipotential location cannot be exposed to dangerous potential difference, in particular, where a conductive floor insulated from Earth is connected to the earth-free equipotential bonding conductors.

413-06 Protection by electrical separation (see also Regulation 471-12)

413-06-01 Protection by electrical separation shall be afforded by compliance with Regulations 413-06-02 and 413-06-03 and with:

(i) Regulation 413-06-04 for a supply to one item of equipment, or

(ii) Regulation 413-06-05 for a supply to more than one item of equipment.

413-06-02 The source of supply to the circuit shall comply with the following requirements:

(i) it shall be either:

(a) an isolating transformer complying with BS 3535, in which there shall be no connection between the output winding and the body or the protective earthing circuit if any, or

(b) a source of current such as a motor generator with windings providing a degree of safety equivalent to that of the safety isolating transformer referred to above.

(ii) a mobile source of supply fed from a fixed installation shall be selected or installed in accordance with Regulation 413-03 (protection by use of Class II equipment or by equivalent insulation).

(iii) equipment used as a fixed source of supply shall be either:

(a) selected and installed in accordance with Regulation 413-03, or

(b) such that the output is separated from the input and from the enclosure by insulation satisfying the conditions of Regulation 413-03. If such a source supplies several items of equipment, exposed-conductive-parts of that equipment shall not be connected to the exposed-conductive-parts of the source.

(iv) the voltage of the electrically separated circuit shall not exceed 500 V.

413-06-03 The separated circuit shall comply with the following requirements:

(i) no live part of the separated circuit shall be connected at any point to another circuit or to Earth and to avoid the risk of a fault to Earth, particular attention shall be given to the insulation of such parts from Earth, especially for flexible cables and cords

(ii) every part of a flexible cable or cord liable to mechanical damage shall be visible throughout its length

(iii) a separate wiring system shall preferably be used for the separated circuit. Alternatively, multicore cables without metallic sheath, or insulated conductors in insulating conduit shall be used, their rated voltage being not less than the highest voltage likely to occur and each circuit shall be protected against overcurrent

(iv) every live part of each separate circuit shall be electrically separated from all other circuits to a standard not less than that provided between the input and output windings of an isolating transformer to BS 3535. The same standard of electrical separation shall also be provided between live parts of relays, contactors etc. included in the separated circuit and between them and live parts of other circuits.

413-06-04 For a circuit supplying a single item of equipment, no exposed-conductive-part of the separated circuit shall be connected either to the protective conductor of the source or to any exposed-conductive-part of any other circuit.

413-06-05 If precautions are taken to protect the separated circuit from damage and insulation failure, a source of supply complying with Regulation 413-06-02 may supply more than one item of equipment provided that all the following requirements are fulfilled:

(i) every exposed-conductive-part of the separated circuit shall be connected together by an insulated and non-earthed equipotential bonding conductor. Such a conductor shall not be connected to a protective conductor or exposed-conductive-part of any other circuit or to any extraneous-conductive-part

(ii) every socket-outlet shall be provided with a protective conductor contact which shall be connected to the equipotential bonding conductor provided in accordance with item (i)

(iii) every flexible cable of equipment other than Class II equipment shall embody a protective conductor for use as an equipotential bonding conductor

(iv) it shall be verified that, if two faults to exposed-conductive-parts occur and these are fed by conductors of different polarity, an associated protective device will meet the requirements of Regulation 413-02-04

(v) no exposed-conductive-part of the source shall be simultaneously accessible with any exposed-conductive-part in the separated circuit.

CHAPTER 42

PROTECTION AGAINST THERMAL EFFECTS

CONTENTS

421 *(Reserved for future use)*

422 PROTECTION AGAINST FIRE AND HARMFUL THERMAL EFFECTS

423 PROTECTION AGAINST BURNS

CHAPTER 42

421 *(Reserved for future use)*

422 PROTECTION AGAINST FIRE AND HARMFUL THERMAL EFFECTS

422-01-01 Fixed electrical equipment shall be selected and installed so that heat generated thereby does not cause danger or harmful effects to adjacent fixed material or to material which may foreseeably be in proximity to such equipment.

In addition, any relevant installation instruction of the equipment manufacturer shall be observed.

422-01-02 Where fixed electrical equipment is installed having, in normal operation, a surface temperature sufficient to cause a risk of fire or harmful effects to adjacent materials, one or more of the following installation methods shall be adopted:

(i) mounting on a support or within an enclosure which will withstand, without risk of fire or harmful effect, such temperatures as may be generated. The support shall have a low thermal conductance, or

(ii) screening by material which can withstand without risk of fire or harmful effect the heat emitted by the electrical equipment, or

(iii) mounting so as to allow safe dissipation of heat and at a sufficient distance from adjacent material.

422-01-03 Where an arc or high temperature particles may be emitted by fixed equipment one or more of the following installation methods shall be adopted:

(i) total enclosure in arc-resistant material

(ii) screening by arc-resistant material, from materials upon which the emissions could have harmful effects

(iii) mounting so as to allow safe extinction of the emissions at a sufficient distance from material upon which the emissions could have harmful effects.

422-01-04 Every termination of live conductors or joint between them shall be contained within an enclosure selected in accordance with Regulation 526-03-02.

422-01-05 Where electrical equipment in a single location contains, in total, flammable liquid in excess of 25 litres, adequate precautions shall be taken to prevent the spread of burning liquid, flame and the products of combustion.

423 PROTECTION AGAINST BURNS

423-01-01 Excepting equipment for which a British Standard specifies a limiting temperature, an accessible part of fixed electrical equipment within arm's reach shall not attain a temperature in excess of the appropriate limit stated in Table 42A. Each such part of the fixed installation likely to attain under normal load conditions, even for a short period, a temperature exceeding the appropriate limit in Table 42A shall be guarded so as to prevent accidental contact.

TABLE 42A

The temperature limit under normal load conditions for an accessible part of equipment within arm's reach

Part	Material of accessible surface	Maximum temperature ^{o}C
A hand-held means of operation	Metallic Non-metallic	55 65
A part intended to be touched but not hand-held	Metallic Non-metallic	70 80
A part which need not be touched for normal operation	Metallic Non-metallic	80 90

CHAPTER 43

PROTECTION AGAINST OVERCURRENT

CONTENTS

431	GENERAL
432	NATURE OF PROTECTIVE DEVICES
432-01	**General**
432-02	**Protection against both overload current and fault current**
432-03	**Protection against overload current only**
432-04	**Protection against fault current only**
433	PROTECTION AGAINST OVERLOAD CURRENT
433-01	**General**
433-02	**Co-ordination between conductor and protective device**
433-03	**Overload protection of conductors in parallel**
434	PROTECTION AGAINST FAULT CURRENT
434-01	**General**
434-02	**Determination of prospective fault current**
434-03	**Characteristics of a fault current protective device**
434-04	**Fault current protection of conductors in parallel**
435	CO-ORDINATION OF OVERLOAD CURRENT AND FAULT CURRENT PROTECTION
436	LIMITATION OF OVERCURRENT BY CHARACTERISTICS OF SUPPLY

CHAPTER 43

PROTECTION AGAINST OVERCURRENT

431 GENERAL

431-01-01 Except where the overcurrent is limited in accordance with Section 436, every live conductor shall be protected by one or more devices for automatic interruption of the supply in the event of overload current (Section 433) and fault current (Section 434), in accordance with Section 473.

431-01-02 The protection against overload current and the protection against fault current shall be co-ordinated in accordance with Section 435.

431-01-03 No fault current shall be allowed to persist indefinitely.

432 NATURE OF PROTECTIVE DEVICES

432-01 **General**

432-01-01 A protective device shall be of the appropriate type indicated in Regulations 432-02 to 432-04.

432-02 **Protection against both overload current and fault current**

432-02-01 Except as permitted by Regulation 434-03-01, a device or devices providing protection against both overload current and fault current shall be capable of breaking and, for a circuit-breaker, making any overcurrent up to and including the prospective fault current at the point where the device is installed. In addition, every device or devices shall satisfy the requirements of Section 433 and Regulations 434-01-01, 434-03-01, 434-03-02 and for conductors in parallel, 434-04-01.

432-03 **Protection against overload current only**

432-03-01 A device providing protection against overload current shall satisfy the requirements of Section 433. Such a device may have a breaking capacity below the value of the prospective fault current at the point where the device is installed.

432-04 **Protection against fault current only**

432-04-01 Except as permitted by Regulation 434-03-01 a device providing protection against fault current shall satisfy the requirements of Section 434. Such a device shall be capable of breaking and, for a circuit-breaker, making any fault current up to and including the prospective fault current.

433 PROTECTION AGAINST OVERLOAD CURRENT

433-01 **General**

433-01-01 A protective device shall be provided to break any overload current flowing in the circuit conductors before such a current causes a temperature rise detrimental to insulation, joints, terminations, or the surroundings of the conductors. Every circuit shall be designed so that a small overload of long duration is unlikely to occur.

433-02 **Co-ordination between conductor and protective device**

433-02-01 The characteristics of each protective device shall satisfy the following conditions:

  (i) its nominal current or current setting (I_n) is not less than the design current (I_b) of the circuit, and

  (ii) its nominal current or current setting (I_n) does not exceed the lowest of the current-carrying capacities (I_z) of any of the conductors of the circuit, and

  (iii) the current causing effective operation of the protective device (I_2) does not exceed 1.45 times the lowest of the current-carrying capacities (I_z) of any of the conductors of the circuit.

433-02-02 Where the device is a general purpose type (gG) fuse to BS 88 Part 2, a fuse to BS 88 Part 6, a fuse to BS 1361 or a circuit-breaker to BS 3871 Part 1, compliance with condition (ii) also results in compliance with condition (iii).

433-02-03 Where the device is a semi-enclosed fuse to BS 3036 compliance with condition (iii) is afforded if its nominal current (I_n) does not exceed 0.725 times the current-carrying capacity of the lowest rated conductor in the circuit protected.

433-03 Overload protection of conductors in parallel

433-03-01 Except for a ring final circuit where a single device provides protection against overload current for conductors in parallel the value of I_z is the sum of the current-carrying capacities of those conductors, which shall be of the same construction, cross-sectional area, length and disposition, have no branch circuits throughout their length and be arranged so as to carry substantially equal currents.

434 PROTECTION AGAINST FAULT CURRENT

434-01 General

434-01-01 A protective device shall be provided in a circuit to break any fault current flowing in conductors of that same circuit before such current causes danger due to thermal or mechanical effects produced in those conductors or the associated connections.

The nominal current of such a protective device may be greater than the current-carrying capacity of the conductor being protected.

434-02 Determination of prospective fault current

434-02-01 The prospective fault current, under both short-circuit and earth fault conditions, at every relevant point of the complete installation shall be determined. This shall be done by either calculation or measurement.

434-03 Characteristics of a fault current protective device

434-03-01 Except where the following paragraph applies, the breaking capacity rating of each device shall be not less than the prospective short-circuit current or earth fault current at the point at which the device is installed.

A lower breaking capacity is permitted if another protective device or devices having the necessary breaking capacity is installed on the supply side. In this situation the characteristics of the devices shall be co-ordinated so that the energy let-through of these devices does not exceed that which can be withstood, without damage, by the device or devices on the load side.

434-03-02 Except as follows (see Regulations 434-03-03 and 434-04-01), where an overload protective device complying with Section 433 is to provide short-circuit current protection and/or earth fault current protection and has a rated breaking capacity not less than the value of the prospective fault current at its point of installation, it may be assumed that the Regulations are satisfied as regards fault current protection of the conductors on the load side of that point.

The validity of the assumption shall be checked, where there is doubt, for conductors in parallel and for certain types of circuit-breaker e.g. non-current-limiting types.

434-03-03 Where a protective device is provided for fault current protection only, the clearance time of the device under both short-circuit and earth fault conditions shall not result in the admissible limiting temperature of any live conductor being exceeded.

The time t, in which a given fault current will raise the live conductors from the highest permissible temperature in normal duty to the limiting temperature, can, as an approximation, be calculated from the formula:

$$t = \frac{k^2 S^2}{I^2}$$

Where:
- t   is the duration in seconds,
- S   is the nominal cross-sectional area of conductor in mm^2
- I   is the value of fault current in amperes, expressed for a.c. as the r.m.s. value, due account being taken of the current limiting effect of the circuit impedances,
- k   is a factor taking account of the resistivity and temperature coefficient and heat capacity of the conductor material, and the appropriate initial and final temperatures. For the common materials indicated in Table 43A, the k factor shall be as shown.

If greater accuracy is required, the calculation shall be made in accordance with the method given in BS 7454.

Other values of k may be used where the initial temperature is lower than the appropriate value stated in the table, i.e., where the conductor concerned is intended to carry a current less than its current-carrying capacity in normal service. For materials other than those mentioned in Table 43A, values of k shall be calculated by an appropriate method, and the values of temperature to be assumed shall be in accordance with the recommendations of the cable manufacturer.

For very short durations (less than 0.1 s) where asymmetry of the current is of importance and for current limiting devices, the value of $k^2 S^2$ for the cable shall be greater than the value of let-through energy ($I^2 t$) of the device as quoted by the manufacturer of the device.

TABLE 43A

Values of k for common materials, for calculation of the effects of fault current

This data is applicable only for disconnection times up to 5 seconds.
For longer times the cable manufacturer shall be consulted.

Conductor material	Insulation material	Assumed initial temperature °C	Limiting final temperature °C	k
Copper	70 °C p.v.c. (General purpose) 85 °C p.v.c. 60 °C rubber 85 °C rubber 90 °C thermosetting Impregnated paper	70 85 60 85 90 80	160/140* 160/140* 200 220 250 160	115/103* 104/90 * 141 134 143 108
Copper	Mineral - plastic covered or exposed to touch - bare and neither exposed to touch nor in contact with combustible materials	70 (sheath) 105 (sheath)	160 250	115 135
Aluminium	70 °C p.v.c. (General purpose) 85 °C p.v.c. 60 °C rubber 85 °C rubber 90 °C thermosetting Impregnated paper	70 85 60 85 90 80	160/140* 160/140* 200 220 250 160	76/68* 69/60* 93 89 94 71

* Where two values of limiting final temperature and of k are given the lower value relates to cables having conductors of greater than 300 mm cross-sectional area.

434-04 Fault current protection of conductors in parallel

434-04-01 Except for a ring final circuit, where a single device provides protection against fault current for conductors in parallel, those conductors shall be of the same construction, cross-sectional area, length and disposition, have no branch circuits throughout their length and be arranged so as to carry substantially equal currents. Compliance with Regulation 434-03-03 shall be verified by calculation, account being taken of the conditions that would occur in the event of a fault which does not affect all of the conductors.

435 CO-ORDINATION OF OVERLOAD CURRENT AND FAULT CURRENT PROTECTION

435-01-01 The characteristics of each device for overload current protection and for fault current protection shall be co-ordinated so that the energy let-through by the fault current protective device does not exceed that which can be withstood without damage by the overload current protective device.

For a circuit incorporating a motor starter, this Regulation does not preclude the type of co-ordination described in BS 4941, in respect of which the advice of the manufacturer of the starter shall be sought.

436 LIMITATION OF OVERCURRENT BY CHARACTERISTICS OF SUPPLY

436-01-01 A conductor shall be considered to be protected against overcurrent where its current-carrying capacity is greater than the current which can be supplied by the source.

CHAPTER 44

(Reserved for future use)

CHAPTER 45

PROTECTION AGAINST UNDERVOLTAGE

451   GENERAL

451-01-01 Suitable precautions shall be taken where a reduction in voltage, or loss and subsequent restoration of voltage, could cause danger. Provisions for a circuit supplying a motor shall comply with Regulation 552-01-03.

451-01-02 Where current-using equipment or any other part of the installation may be damaged by a drop in voltage and it is verified that such damage is unlikely to cause danger, one of the following arrangements shall be adopted:

    (i)   suitable precautions against the damage foreseen shall be provided, or

    (ii)   it shall be verified, in consultation with the person or body responsible for the operation and maintenance of the installation, that the damage foreseen is an acceptable risk.

451-01-03 A suitable time delay may be incorporated in the operation of an undervoltage protective device if the operation of the equipment to which the protection relates allows without danger a brief reduction or loss of voltage.

451-01-04 Any delay in the opening or reclosing of a contactor shall not impede instantaneous disconnection by a control device or a protective device.

451-01-05 The characteristics of an undervoltage protective device shall be compatible with the requirements for starting and use of the equipment to which the protection relates, as stated in the appropriate British Standard.

451-01-06 Where the reclosure of a protective device is likely to cause danger, the reclosure shall not be automatic.

CHAPTER 46

ISOLATION AND SWITCHING

CONTENTS

460   GENERAL

461   ISOLATION

462   SWITCHING OFF FOR MECHANICAL MAINTENANCE

463   EMERGENCY SWITCHING

CHAPTER 46

ISOLATION AND SWITCHING

460   GENERAL

460-01-01 A means shall be provided for non-automatic isolation and switching to prevent or remove hazards associated with the electrical installation or electrically powered equipment and machines. Such means shall comply with the appropriate requirements of this Chapter and of Sections 476 and 537.

460-01-02   A main linked switch or circuit-breaker (linked) shall be provided as near as practicable to the origin of every installation as a means of interrupting the supply on load and as a means of isolation.

This main switch or circuit-breaker shall interrupt both live conductors of a single-phase supply in any system and for three-phase supplies the requirements of Regulations 461-01-01 and 461-01-03 shall apply. For d.c. supplies, all poles shall be interrupted.

Where an installation is supplied from more than one source, a main switch shall be provided for each source of supply and a durable warning notice shall be permanently fixed in such a position that any person seeking to operate any of these main switches will be warned of the need to operate all such switches to achieve isolation of the installation. Alternatively, a suitable interlock system shall be provided.

460-01-03   Except as provided by Regulation 460-01-04, neither an isolator (disconnector) nor a switch shall break a protective conductor or a PEN conductor.

460-01-04   Where an installation is supplied from more than one source of energy, one of which requires a means of earthing independent of the means of earthing of other sources and it is necessary to ensure that not more than one means of earthing is applied at any time, a switch may be inserted in the connection between the neutral point and the means of earthing, provided that the switch is a linked switch arranged to disconnect and connect the earthing conductor for the appropriate source, at substantially the same time as the related live conductors.

461   ISOLATION

461-01-01   Every circuit shall be provided with means of isolation in accordance with Regulation 461-01-02 and 461-01-03. It is permissible to isolate a group of circuits by a common means, due consideration being given to service conditions.

461-01-02   Except at the origin of the installation where Regulation 460-01-02 applies, in an installation forming part of a TN-S or TN-C-S system, provision shall be made for isolation from all phase conductors.

461-01-03   In an installation forming part of a TT or IT system provision shall be made for isolation from all live conductors.

461-01-04   Suitable provision shall be made so that precautions can be taken to prevent any equipment from being inadvertently or unintentionally energised.

461-01-05   Where an item of equipment or enclosure contains live parts that are not capable of being isolated by a single device, a durable warning notice shall be permanently fixed in such a position that any person before gaining access to live parts will be warned of the need to use the appropriate isolating devices, unless an interlocking arrangement is provided so that all the circuits concerned are isolated before access is gained.

461-01-06   Where necessary to prevent danger, adequate means shall be provided for the discharge of capacitive or inductive electrical energy.

461-01-07   Each device used for isolation shall be clearly identified by durable marking to indicate the installation or circuit which it isolates.

462   SWITCHING OFF FOR MECHANICAL MAINTENANCE

462-01-01   A means of switching off for mechanical maintenance shall be provided where mechanical maintenance may involve a risk of burns or a risk of injury from mechanical movement.

462-01-02   Each device for switching off for mechanical maintenance shall be suitably located, and identified by durable marking where necessary.

462-01-03   Except where the means of switching off is continuously under the control of any person performing such maintenance, suitable provision shall be made so that precautions can be taken to prevent any equipment from becoming unintentionally or inadvertently reactivated during mechanical maintenance.

463   EMERGENCY SWITCHING

463-01-01   A means of emergency switching shall be provided for every part of an installation which it may be necessary to cut off rapidly from the supply in order to prevent or remove danger.

The means shall interrupt both live conductors of a single phase supply in any system and for a three phase supply the requirements of Regulations 461-01-02 and 461-01-03 shall apply. For a d.c. supply, all poles shall be interrupted.

Except as provided in Regulation 460-01-02, where a risk of electric shock is involved the means shall break all live conductors.

463-01-02 A means of emergency switching shall act as directly as possible on the appropriate supply conductors, and shall be such that only a single initiative action is required.

463-01-03 The arrangement of emergency switching shall be such that its operation does not introduce a further hazard or interfere with the complete operation necessary to remove the hazard.

463-01-04 Each device for emergency switching shall be readily accessible and durably marked.

463-01-05 A means of emergency stopping shall be provided where mechanical movement of electrically actuated equipment may give rise to danger.

CHAPTER 47

APPLICATION OF PROTECTIVE MEASURES FOR SAFETY

CONTENTS

470	GENERAL
471	PROTECTION AGAINST ELECTRIC SHOCK
471-01	**General**
471-02 and **471-03**	*Protection against both direct and indirect contact*
471-02	**Protection by SELV**
471-03	**Protection by limitation of discharge of energy**
471-04 to **471-07**	*Protection against direct contact*
471-04	**Protection by insulation of live parts**
471-05	**Protection by barriers or enclosures**
471-06	**Protection by obstacles**
471-07	**Protection by placing out of reach**
471-08 to **471-12**	*Protection against indirect contact*
471-08	**Protection by earthed equipotential bonding and automatic disconnection of supply**
471-09	**Protection by Class II equipment or equivalent insulation**
471-10	**Protection by non-conducting location**
471-11	**Protection by earth-free local equipotential bonding**
471-12	**Protection by electrical separation**
471-13	**Special provisions and exemptions**
471-14	**Functional extra-low voltage systems**
471-15	**Automatic disconnection and reduced low voltage systems**

471-16	**Supplies for portable equipment outdoors**
472	*(Reserved for future use)*
473	PROTECTION AGAINST OVERCURRENT
473-01	**Protection against overload**
473-01-01 and 473-01-02	*Position of devices for overload protection*
473-01-03 and 473-01-04	*Conditions for omission of devices for overload protection*
473-01-05	*Overload protective devices in IT systems*
473-02	**Protection against fault current**
473-02-01 to 473-02-03	*Position of devices for fault current protection*
473-02-04	*Conditions for omission of devices for fault current protection*
473-03	**Protection according to the nature of circuit and distribution systems**
473-03-01 and 473-03-02	*Phase conductors*
473-03-03 and 473-03-04	*Neutral conductor - TN or TT systems*
473-03-05	*Neutral conductor - IT systems*
474	*(Reserved for future use)*
475	*(Reserved for future use)*
476	ISOLATION AND SWITCHING
476-01	**General**
476-02	**Isolation**
476-03	**Emergency switching**

CHAPTER 47

APPLICATION OF PROTECTIVE MEASURES FOR SAFETY

470 GENERAL

470-01-01 Protective measures shall be applied in every installation or part of an installation, or to equipment, as required by:

Section 471 - Protection against electric shock.
Section 473 - Protection against overcurrent.
Section 476 - Isolation and switching.

470-01-02 Precautions shall be taken so that no detrimental influence can occur between various protective measures in the same installation or part of an installation.

471 PROTECTION AGAINST ELECTRIC SHOCK

471-01 General

471-01-01 The application of the various protective measures described in Chapter 41 (relevant Regulation numbers indicated below in brackets) is qualified as follows.

Regulations 471-02 to 471-14, apply, except where otherwise stated, to normal dry conditions where a person can be assumed to have conventional normal body resistance and has no contact directly with earth potential.

Regulations 471-15 and 471-16 prescribe additional particular requirements for installations and locations where the risk of electric shock is increased by reduction in body resistance or by contact with earth potential.

Protection against both direct and indirect contact

471-02 Protection by SELV (Regulation 411-02)

471-02-01 This measure is generally applicable except that for some installations and locations of increased shock risk:
  (i)   it is the only measure against electric shock permitted, and

  (ii) a reduction in the nominal voltage is prescribed, and

  (iii) protection against direct contact, as prescribed in Regulation 411-02-09, shall be provided irrespective of the nominal voltage.

For these installations and locations, see the particular requirements of Part 6 of the Regulations.

471-03 Protection by limitation of discharge of energy (Regulation 411-04)

471-03-01 This measure shall be applied only to an individual item of current-using equipment complying with an appropriate British Standard, where the equipment incorporates means of limiting to a safe value the current that can flow from the equipment through the body of a person or livestock. The application of this measure may be extended to a part of an installation derived from such items of equipment, where the British Standard concerned provides specifically for this, e.g. to electric fences supplied from electric fence controllers complying with BS 2632.

Protection against direct contact

471-04 Protection by insulation of live parts (Regulation 412-02)

471-04-01 This measure relates to basic insulation, and is intended to prevent contact with a live part. It is generally applicable for protection against direct contact, in conjunction with a measure for protection against indirect contact.

471-05 Protection by barriers or enclosures (Regulation 412-03)

471-05-01 This measure is intended to prevent or deter any contact with a live part. It is generally applicable for protection against direct contact, in conjunction with a measure for protection against indirect contact.

471-05-02 The exception in Regulation 412-03-01 allowing for an opening larger than IP 2X in a barrier or enclosure shall be applied only to an item of equipment or accessory complying with a British Standard where compliance with the generality of Regulation 412-03-01 is impracticable by reason of the function of the item, e.g. to a lampholder complying with BS 5042 or other appropriate Standard. Wherever that exception is used, the opening shall be as small as is consistent with the requirement for proper functioning and for replacement of a part.

471-06 Protection by obstacles (Regulation 412-04)

471-06-01 This measure is intended to prevent unintentional contact with a live part, but not intentional contact by deliberate circumvention of the obstacle. The application shall be limited to protection against direct contact and in an area accessible only to skilled persons, or to instructed persons under the direct supervision of a skilled person.

For some installations and locations of increased shock risk this protective measure shall not be used. See the particular requirements of Part 6 of the Regulations.

471-07 Protection by placing out of reach (Regulation 412-05)

471-07-01 This measure is intended to prevent unintentional contact with a live part and shall be applied for protection against direct contact. The application of the provisions of Regulations 412-05-02 to 412-05-04 shall be limited to each location accessible only to skilled persons, or instructed persons under the direct supervision of a skilled person.

For some installations and locations of increased shock risk this protective measure shall not be used. See the particular requirements of Part 6 of the Regulations.

Protection against indirect contact

471-08 Protection by earthed equipotential bonding and automatic disconnection of supply (Regulation 413-02)

471-08-01 This measure is generally applicable, and is intended to prevent the occurrence of a voltage of such magnitude and duration between simultaneously accessible conductive parts that danger could arise.

For some installations and locations of increased shock risk:

(i) automatic disconnection of supply shall be by means of a residual current device having a rated residual current not exceeding 30 mA,

(ii) supplementary equipotential bonding shall be applied, and

(iii) maximum fault clearance times shall be reduced.

For some special installations and locations, see the particular requirements of Part 6 of the Regulations.

471-08-02 For an installation which is part of a TN system, the limiting values of earth fault loop impedance and of circuit protective conductor impedance specified by Regulations 413-02-08 and 413-02-10 to 413-02-16 are applicable only where the exposed-conductive-parts of the equipment concerned and any extraneous-conductive-parts are situated within the earthed equipotential zone (see also Regulation 413-02-02).

Where the disconnection times specified by Regulation 413-02-08 cannot be met by the use of an overcurrent protective device, Regulation 413-02-15 shall be applied.

471-08-03 Where a circuit supplies fixed equipment outside the earthed equipotential zone and the equipment has exposed-conductive-parts which may be touched by a person in contact directly with the general mass of Earth, the earth fault loop impedance shall be such that disconnection occurs within the time stated in Table 41A.

471-08-04 A socket-outlet circuit rated 32 A or less which may reasonably be expected to supply portable equipment for use outdoors shall comply with Regulation 471-16-01.

471-08-05 A circuit supplying portable equipment for use outdoors connected by means of a flexible cable or cord having a current-carrying capacity of 32 A or less, other than through a socket-outlet, shall comply with Regulation 471-16-02.

471-08-06 Where the measure is used in an installation forming part of a TT system, every socket-outlet circuit shall be protected by a residual current device and shall comply with Regulation 413-02-16.

471-08-07 Automatic disconnection using a residual current device shall not be applied to a circuit incorporating a PEN conductor.

471-08-08 In every installation which provides for protection against indirect contact by automatic disconnection of supply, a circuit protective conductor shall be run to and terminated at each point in wiring and at each accessory except a lampholder having no exposed-conductive-parts and suspended from such a point.

471-09 Protection by the use of Class II equipment or equivalent insulation
(Regulation 413-03)

471-09-01 This measure is intended to prevent the appearance of a dangerous voltage on the exposed metalwork of electrical equipment through a fault in the basic insulation. It is generally applicable to an item of equipment, either by the selection of equipment complying with an appropriate British Standard where that Standard provides for the use of Class II construction or total insulation, or by the application of suitable supplementary insulation during erection.

471-09-02 Where a circuit supplies items of Class II equipment, a circuit protective conductor shall be run to and terminated at each point in wiring and at each accessory except a lampholder having no exposed-conductive-parts and suspended from such a point. This requirement need not be observed where Regulation 471-09-03 applies.

Exposed metalwork of Class II equipment shall be mounted so that it is not in electrical contact with any part of the installation connected to a protective conductor. Such a contact might impair the Class II protection provided by the equipment specification. In case of doubt the appropriate British Standard for the equipment, or the manufacturer, shall be consulted.

471-09-03 Where this measure is to be used as a sole means of protection against indirect contact (i.e. where a whole installation or circuit is intended to consist entirely of Class II equipment or the equivalent), it shall be verified that the installation or circuit concerned will be under effective supervision in normal use so that no change is made that would impair the effectiveness of the Class II or equivalent insulation. The measure shall not therefore be applied to any circuit which includes a socket-outlet or where a user may change items of equipment without authorisation.

471-09-04 Cable having a non-metallic sheath or a non-metallic enclosure shall not be described as being of Class II construction. However, the use of such cable installed in accordance with Chapter 52 shall be deemed to afford satisfactory protection against direct and indirect contact.

471-10 Protection by non-conducting location (Regulation 413-04)

471-10-01 This measure is intended to prevent simultaneous contact with parts which may be at different potentials through failure of the basic insulation of live parts.

This measure is not recognised in the Regulations for general use and shall be applied only in special situations which are under effective supervision.

This measure shall not be used in certain installations and locations of increased shock risk covered by Part 6 of the Regulations.

471-11 Protection by earth-free local equipotential bonding (Regulation 413-05)

471-11-01 This measure is intended to prevent the appearance of a dangerous voltage between simultaneously accessible parts in the event of failure of the basic insulation. It shall be applied under effective supervision only in special situations which are Earth free. Where this measure is applied, a warning notice complying with Regulation 514-13-02 shall be fixed in a prominent position adjacent to every point of access to the location concerned.

For some installations and locations of increased shock risk this measure shall not be used. See the particular requirements of Part 6 of the Regulations.

471-12 Protection by electrical separation (Regulation 413-06)

471-12-01 This measure is intended, in an individual circuit, to prevent shock current through contact with exposed-conductive-parts which might be energised by a fault in the basic insulation of that circuit. It may be applied to the supply of any individual item of equipment by means of a transformer complying with BS 3535 the secondary of which is not earthed, or a source affording equivalent safety. Its use to supply several items of equipment from a single separated source is recognised in the Regulations only for special situations under effective supervision, where specified by a suitably qualified electrical engineer. Where the measure is used to supply several items of equipment from a single source, a warning notice complying with Regulation 514-13-02 shall be fixed in a prominent position adjacent to every point of access to the location concerned.

471-13 Special provisions and exemptions

471-13-01 For an area to which only skilled persons, or instructed persons directly supervised by a skilled person, have access it is sufficient to provide against unintentional contact with live parts by the use of an obstacle in accordance with Regulations 412-04 and 471-06, or by placing each live part out of reach in accordance with Regulations 412-05 and 471-07, subject also to Regulations 471-13-02 and 471-13-03.

471-13-02 The dimensions of each passageway and working platform for an open type switchboard and other equipment having dangerous exposed live parts shall be adequate such as to allow persons, without hazard, to:

(i) operate and maintain the equipment,
(ii) pass one another as necessary with ease, and
(iii) back away from the equipment.

471-13-03 Areas reserved for skilled or instructed persons shall be clearly and visibly indicated by suitable warning signs.

471-13-04 It is permissible to dispense with protective measures against indirect contact in the following instances:

(i) overhead line insulator brackets and metal parts connected to them if such parts are not situated within arm's reach

(ii) steel reinforced concrete poles in which the steel reinforcement is not accessible

(iii) exposed-conductive-parts which, owing to their reduced dimensions or their disposition cannot be gripped or cannot be contacted by a major surface of the human body, provided that connection of these parts to a protective conductor cannot readily be made or cannot be reliably maintained. This dispensation includes small isolated metal parts such as bolts, rivets, nameplates not exceeding 50 mm x 50 mm and cable clips

(iv) fixing screws for non-metallic accessories provided that there is no appreciable risk of the screws coming into contact with live parts

(v) inaccessible lengths of metal conduit not exceeding 150 mm

(vi) metal enclosures mechanically protecting equipment complying with Regulations 413-03-01 to 413-03-09 and 471-09-01

(vii) unearthed street furniture supplied from an overhead line and inaccessible in normal use.

471-14 Functional extra-low voltage systems

471-14-01 Where for functional reasons extra-low voltage is used but not all the requirements of Regulation 411-02 regarding SELV are fulfilled, the appropriate measures described in Regulations 471-14-02 to 471-14-06 shall be taken in order to provide protection against electric shock. Systems employing these measures are termed Functional extra-low voltage systems.

471-14-02 If the extra-low voltage system complies with the requirements of Section 411 for SELV except that live or exposed-conductive-parts are connected to Earth or to the protective conductors of other systems (see Regulations 411-02-05 and 411-02-07) protection against direct contact shall be provided by one or more of the following:

(i) enclosures giving protection at least equivalent to IP 2X (see BS 5490)

(ii) insulation capable of withstanding a test voltage of 500 V r.m.s. a.c. for 60 s.

Such a system shall be deemed to afford protection against indirect contact.

This requirement shall not exclude the installation or the use without supplementary protection of equipment conforming to the relevant British Standard, e.g. BS 415, providing an equivalent degree of safety.

471-14-03 If the extra-low voltage system does not comply with the requirements of Section 411 for SELV in some respect other than that specified in Regulation 471-14-02, protection against direct contact shall be provided by one or more of the following:

(i) barriers or enclosures according to Regulation 412-03

(ii) insulation corresponding to the minimum test voltage required for the primary circuit.

In addition, protection against indirect contact shall be provided in accordance with Regulations 471-14-04 or 471-14-05.

Where the extra-low voltage circuit is used to supply equipment whose insulation does not comply with the minimum test voltage required for the primary circuit, the accessible insulation of that equipment shall be reinforced during erection to withstand a test voltage of 1500 V r.m.s. a.c. for 60 s.

471-14-04   If the primary circuit of the functional extra-low voltage source is protected by automatic disconnection, exposed-conductive-parts of equipment in the functional extra-low voltage system shall be connected to the protective conductor of the primary circuit. This shall not exclude the possibility of connecting a conductor of the functional extra-low voltage circuit to the protective conductor of the primary circuit.

471-14-05   If the primary circuit of the functional extra-low voltage source is protected by electrical separation (Regulation 413-06), the exposed-conductive-parts of equipment in the functional extra-low voltage circuit shall be connected to the non-earthed protective conductor of the primary circuit. This latter requirement shall be deemed not to contravene Regulations 413-06-04 and 413-06-05, the combination of the electrically separated circuit and the extra-low voltage circuit being regarded as one electrically separated circuit.

471-14-06   Every socket-outlet and luminaire supporting coupler in a functional extra-low voltage system shall require the use of a plug which is not dimensionally compatible with those used for any other system in use in the same premises.

471-15   Automatic disconnection and reduced low voltage systems

471-15-01   Where for functional reasons the use of extra-low voltage is impracticable and there is no requirement for the use of SELV, a reduced low voltage system may be used as specified in Regulations 471-15-02 to 471-15-07.

471-15-02   The nominal voltage of the reduced low voltage circuits shall not exceed 110 V r.m.s. a.c. between phases (three-phase 63.5 V to earthed neutral, single-phase 55 V to earthed midpoint).

471-15-03   The source of supply to a reduced low voltage circuit shall be one of the following:

(i) a double wound isolating transformer complying with BS 3535: Part 2

(ii) a motor generator having windings providing isolation equivalent to that provided by the windings of an isolating transformer

(iii) a source independent of other supplies, e.g. an engine driven generator.

471-15-04   The neutral (star) point of the secondary windings of three-phase transformers and generators, or the midpoint of the secondary windings of single-phase transformers and generators, shall be connected to Earth.

471-15-05   Protection against direct contact shall be provided by insulation in accordance with Regulations 412-02 and 471-04 or by a barrier or enclosure in accordance with Regulations 412-03 and 471-05.

471-15-06   Protection against indirect contact by automatic disconnection shall be provided by means of an overcurrent protective device in each phase conductor or by a residual current device, and all exposed conductive parts of the reduced low voltage system shall be connected to Earth. The earth fault loop impedance at every point of utilisation, including socket-outlets, shall be such that the disconnection time does not exceed 5 s.

Where a circuit-breaker is used, the maximum value of earth fault loop impedance (Z_s) shall be determined by the formula of Regulation 413-02-08. Alternatively the values specified in Table 471A may be used instead of calculation for the nominal voltages to Earth (U_o) and for the types and ratings of circuit-breaker listed therein.

Where a fuse is used, maximum values of earth fault loop impedance (Z_s) corresponding to a disconnection time of 5 s are stated in Table 471A for nominal voltages to Earth (U_o) of 55 V and 63.5 V.

For types and rated currents of fuses other than those mentioned in Table 471A, reference should be made to the appropriate British Standard, to determine the value of I_a for compliance with Regulation 413-02-08, according to the appropriate value of the nominal voltage to Earth (U_o).

Where a residual current device is used, the product of the rated residual operating current in amperes and the earth fault loop impedance in ohms shall not exceed 50.

TABLE 471A

Maximum earth fault loop impedance (Z_s ohms) for a disconnection time 5 s and U_o 55 V (single-phase) and 63.5 V (three-phase) - see Regulations 471-15-02 and 471-15-06

	Miniature circuit breakers to BS 3871 m.c.b. type								General Purpose (gG) fuses to BS88 Parts	
	1		B		2		C and 3		2 and 6	
U_o (Volts)	55	63.5	55	63.5	55	63.5	55	63.5	55	63.5
Rating Amperes										
5	2.76	3.18	2.20	2.54	1.58	1.82	1.10	1.28	-	-
6	2.30	2.65	1.83	2.12	1.32	1.52	0.92	1.07	3.20	3.70
10	1.38	1.59	1.10	1.27	0.79	0.91	0.55	0.64	1.77	2.05
15	0.92	1.06	0.73	0.85	0.53	0.61	0.37	0.43	-	-
16	0.86	0.99	0.69	0.79	0.49	0.57	0.34	0.40	1.00	1.15
20	0.69	0.80	0.55	0.64	0.40	0.46	0.28	0.32	0.69	0.80
25	0.55	0.64	0.44	0.51	0.32	0.36	0.22	0.26	0.55	0.63
30	0.46	0.53	0.37	0.42	0.26	0.30	0.18	0.21	-	-
32	0.43	0.50	0.34	0.40	0.25	0.28	0.17	0.20	0.44	0.51
40	0.35	0.40	0.28	0.32	0.20	0.23	0.14	0.16	0.32	0.37
50	0.28	0.32	0.22	0.25	0.16	0.18	0.11	0.13	0.25	0.29
63	0.22	0.25	0.17	0.20	0.13	0.14	0.09	0.10	0.20	0.23
80	0.17	0.20	0.14	0.16	0.10	0.11	0.07	0.08	0.14	0.16
100	0.14	0.16	0.11	0.13	0.08	0.09	0.05	0.06	0.10	0.12
I_n	13.8 I_n	15.9 I_n	11 I_n	12.7 I_n	7.9 I_n	9.1 I_n	5.5 I_n	6.4 I_n		

471-15-07 Every plug, socket-outlet and cable coupler of a reduced low voltage system shall have a protective conductor contact and shall not be dimensionally compatible with any plug, socket-outlet or cable coupler for use at any other voltage or frequency in the same installation.

471-16 Supplies for portable equipment outdoors

471-16-01 A socket-outlet rated at 32 A or less which may reasonably be expected to supply portable equipment for use outdoors shall be provided with supplementary protection to reduce the risk associated with direct contact by means of a residual current device having the characteristics specified in Regulation 412-06-02(ii).

This Regulation does not apply to a socket-outlet supplied by a circuit incorporating one or more of the protective measures specified in Items (i) to (iii) below and complying with the Regulations indicated:

(i) protection by SELV (see Regulations 411-02 and 471-02)

(ii) protection by electrical separation (see Regulations 413-06 and 471-12)

(iii) protection by automatic disconnection and reduced low voltage systems (see Regulation 471-15).

471-16-02 Except where one or more of the protective measures specified in Items (i) to (iii) of Regulation 471-16-01 are applied in compliance with the corresponding Regulations stated therein, a circuit supplying portable equipment for use outdoors, connected other than through a socket-outlet, by means of flexible cable or cord having a current-carrying capacity of 32 A or less, shall be provided with supplementary protection to reduce the risk associated with direct contact, by means of residual current device having the characteristics specified in Regulation 412-06-02(ii).

472 *(Reserved for future use).*

473 **PROTECTION AGAINST OVERCURRENT**

473-01 **Protection against overload**

Position of devices for overload protection

473-01-01 A device for protection against overload shall be placed at the point where a reduction occurs in the value of current-carrying capacity of the conductors of the installation due to a change in cross-sectional area, method of installation, type of cable or conductor, or in environmental conditions. This requirement does not apply where the arrangements mentioned in Regulation 473-01-02 are adopted, and no overload protective device need be provided where Regulation 473-01-04 applies.

473-01-02 A device protecting a conductor against overload may be placed along the run of that conductor, provided that the part of the run between the point where the value of current-carrying capacity is reduced and the position of the protective device has no branch circuit or outlet for the connection of current-using equipment.

Conditions for omission of devices for overload protection

473-01-03 A device for overload protection shall not interrupt the secondary circuit of a current transformer.

473-01-04 Devices for protection against overload need not be provided:

(i) for a conductor situated on the load side of the point where a reduction occurs in the value of current-carrying capacity, where the conductor is effectively protected against overload by a protective device placed on the supply side of that point

(ii) for a conductor which, because of the characteristics of the load or the supply, is not likely to carry overload current

(iii) in a circuit supplying equipment where unexpected opening of the circuit causes a greater danger than an overload condition

(iv) at the origin of an installation where the supplier provides an overload device and agrees that it affords protection to the part of the installation between the origin and the main distribution point of the installation where further overload protection is provided.

Overload protective devices in IT systems

473-01-05 The provisions of Regulations 473-01-02 and 473-01-04 are applicable to an installation forming part of an IT system only where the conductors concerned are protected by a residual current device, or all the equipment supplied by the circuit concerned (including the conductors) complies with the protective measure described in Regulations 413-03-01 to 413-03-09 (i.e. protection by use of Class II equipment or by equivalent insulation).

473-02 **Protection against fault current**

Position of devices for fault current protection

473-02-01 A device for protection against fault current shall be placed at the point where a reduction occurs in the value of current-carrying capacity of the conductors of the installation such as may be due to a change in cross-sectional area, method of installation, type of cable or conductor, or in environmental conditions. This requirement does not apply where the arrangements mentioned in Regulations 473-02-02 or 473-02-03 are adopted, and no fault current protective device need be provided where Regulation 473-02-04 applies.

473-02-02 The fault current protective device may be placed at a point on the load side of that specified in Regulation 473-02-01 under the following conditions:

Between the point where the value of current-carrying capacity is reduced and the position of the protective device, each conductor shall:

(i) not exceed 3 m in length, and

(ii) be erected in such a manner as to reduce to a minimum the risk of fault current, and

(iii) be erected in such a manner as to reduce to a minimum the risk of fire or danger to persons.

473-02-03 The fault current protective device may be placed at a point other than that specified in Regulation 473-02-01, where a protective device on the supply side of that point has an operating characteristic such that it protects the conductors on the load side of that point against fault current, in accordance with Regulation 434-03-03.

Conditions for omission of devices for fault current protection

473-02-04 A device for protection against fault current may be omitted in the following circumstances provided that the conductor thus not protected against fault current complies with conditions (ii) and (iii) of Regulation 473-02-02:

(i) for a conductor connecting a generator, transformer, rectifier or battery with its control panel where it is protected by a fault current device in the panel

(ii) in a measuring circuit where disconnection could cause danger

(iii) in a circuit supplying equipment where unexpected opening of the circuit causes a greater danger than a fault current condition

(iv) at the origin of an installation where the supplier provides a fault current device and agrees that device affords protection to the parts of the installation between the origin and the main distribution point of the installation where the next step for fault current protection is provided.

473-03 Protection according to the nature of circuits and distribution systems

Phase conductors

473-03-01 A means of detection of overcurrent shall be provided for each phase conductor, and shall cause the disconnection of the conductor in which the overcurrent is detected, but not necessarily the disconnection of other live conductors. Where the disconnection of one phase could cause danger, for example in the supply to three-phase motors, appropriate precautions shall be taken.

473-03-02 In a TT system, for a circuit supplied between phases and in which the neutral conductor is not distributed, overcurrent protection need not be provided for one of the phase conductors, provided that both the following conditions are fulfilled:

(i) there exists, in the same circuit or on the supply side, differential protection intended to cause disconnection of all the phase conductors, and

(ii) the neutral conductor is not distributed from an artificial neutral point of the circuit situated on the load side of that differential protective device.

Neutral conductor - TN or TT Systems

473-03-03 In an installation forming part of TN or TT systems, where the cross-sectional area of the neutral conductor is at least equal or equivalent to that of the phase conductors, it is not necessary to provide overcurrent detection for the neutral conductor or a disconnecting device for this conductor.

473-03-04 In an installation forming part of TN or TT systems where the cross-sectional area of the neutral conductor is less than that of the phase conductors, overcurrent detection shall be provided for the neutral conductor, appropriate to the cross-sectional area of the conductor. This detection shall cause the disconnection of the phase conductors but not necessarily of the neutral conductor.

However, overcurrent detection need not be provided for the neutral conductor if both the following conditions are fulfilled:

(i) the neutral conductor is protected against fault current by the protective device for the phase conductors of the circuit, and

(ii) the maximum current likely to be carried by the neutral conductor in normal service, is significantly less than the value of the current-carrying capacity of that conductor.

Neutral conductor - IT Systems

473-03-05 Where the neutral conductor is distributed then, except where the conditions described in (i) or (ii) below apply, overcurrent detection shall be provided for the neutral conductor of every circuit. Such overcurrent detection shall cause disconnection of all live conductors (including the neutral) of the circuit with which it is associated. The exceptional conditions are:

(i) where the particular neutral conductor is effectively protected against fault current by a protective device placed on the supply side, for example at the origin of the installation, in accordance with Regulation 434-03-01

(ii) where the particular circuit is protected by a residual current operated protective device with a rated residual current not exceeding 0.15 times the current-carrying capacity of the corresponding neutral conductor. This device shall disconnect all the live conductors of the corresponding circuit, including the neutral conductor.

474 *(Reserved for future use)*

475 *(Reserved for future use)*

476 ISOLATION AND SWITCHING

476-01 **General**

476-01-01 Every installation shall be provided with means of isolation and switching complying with Chapter 46, and with the other means of switching for safety required by Regulation 476-01-02. This requirement is satisfied where the supplier provides switchgear complying with Chapter 46 at the origin of the installation and agrees that it may be used as the means of isolation for that part of the installation between the origin and the main distribution point of the installation where the next step for isolation is provided.

Means of electrical switching off for mechanical maintenance, or means of emergency switching, or both, shall be provided for any parts of the installation to which Section 462 or 463 respectively applies.

Where more than one of these functions are to be performed by a common device, the arrangement and characteristics of the device shall satisfy all the requirements of the Regulations for the various functions concerned. Devices for functional switching may serve also for the purposes described above where they satisfy the relevant requirements.

476-01-02 Every circuit and final circuit shall be provided with a means of switching for interrupting the supply on load. A group of circuits may be switched by a common device. Additionally, such means shall be provided for every circuit or other part of the installation which it may be necessary for safety reasons to switch independently of other circuits or other parts of the installation. This Regulation does not apply to short connections between the origin of the installation and the consumer's main switchgear.

476-02 **Isolation**

476-02-01 Where an isolator (disconnector) is to be used in conjunction with a circuit-breaker as a means of isolating main switchgear for maintenance, it shall be interlocked with the circuit-breaker; alternatively, it shall be so placed and/or guarded that it can be operated only by skilled persons.

476-02-02 Where an isolating device for a particular circuit is placed remotely from the equipment to be isolated, provision shall be made so that the means of isolation can be secured in the open position. Where this provision takes the form of a lock or removable handle, the key or handle shall be non-interchangeable with any others used for a similar purpose within the premises.

476-02-03 Every motor circuit shall be provided with an isolating device or devices which shall disconnect the motor and all equipment, including any automatic circuit-breaker, used therewith.

476-02-04 For every electric discharge lighting installation operating normally at an open-circuit voltage exceeding low voltage, one or more of the following means shall be provided for the isolation of every self-contained luminaire, or of every circuit supplying luminaires at a voltage exceeding low voltage:

(i) an interlock on a self-contained luminaire, so arranged that before access can be had to live parts the supply is automatically disconnected, such means being additional to the switch normally used for controlling the circuit

(ii) an effective local means for the isolation of the circuit from the supply, such means being additional to the switch normally used for controlling the circuit

(iii) a switch having a lock or removable handle, or a distribution board which can be locked, in either case complying with Regulation 476-02-02.

476-03   Emergency Switching

476-03-01   For every emergency switching device, account shall be taken of the intended use of the premises so that access to the device is not likely to be impeded in the conditions of emergency foreseen.

476-03-02   A means of emergency stopping shall be provided in every place where an electrically driven machine may give rise to danger. This means shall be readily accessible and easily operated. Where more than one means of starting the machine is provided and danger might be caused by unexpected restarting, means shall be provided to prevent such restarting.

476-03-03   Where additional danger could arise from inappropriate operation of an emergency switching device, the switching device shall be arranged so as to be available for operation by skilled and instructed persons only.

476-03-04   Every fixed or stationary appliance which may give rise to a hazard in normal use and is connected to the supply other than by means of a plug and socket-outlet complying with Regulation 537-05-01 shall be provided with a means of interrupting the supply on load. The operation of the means of interrupting the supply on load shall be so placed as not to put the operator in danger. This means may be incorporated in the appliance or, if separate from the appliance, shall be in a readily accessible position. Where two or more such appliances are installed in the same room one interrupting means may be used to control all the appliances.

476-03-05   A fireman's emergency switch shall be provided in the low voltage circuit supplying:

(i)   exterior electrical installations operating at a voltage exceeding low voltage, and

(ii)   interior discharge lighting installations operating at a voltage exceeding low voltage.

For the purpose of this Regulation, an installation in a covered market, arcade or shopping mall is considered to be an exterior installation. A temporary installation in a permanent building used for exhibitions is considered not to be an exterior installation.

This requirement does not apply to a portable discharge lighting luminaire or to a sign of rating not exceeding 100 W and fed from a readily accessible socket-outlet.

476-03-06   Every exterior installation covered by Regulation 476-03-05 on each single premises shall be wherever practicable controlled by a single fireman's switch. Similarly, every internal installation covered by Regulation 476-03-05 in each single premises shall be controlled by a single fireman's switch independent of the switch for any exterior installation.

476-03-07   Every fireman's emergency switch provided for compliance with Regulation 476-03-05 shall comply with all the relevant requirements of the following items (i) to (iv) and any requirements of the local fire authority:

(i)   for an exterior installation, the switch shall be outside the building and adjacent to the equipment, or alternatively a notice indicating the position of the switch shall be placed adjacent to the equipment and a notice shall be fixed near the switch so as to render it clearly distinguishable

(ii)   for an interior installation, the switch shall be in the main entrance to the building or in another position to be agreed with the local fire authority

(iii)   the switch shall be placed in a conspicuous position, reasonably accessible to firemen and, except where otherwise agreed with the local fire authority, at not more than 2.75 m from the ground or the standing beneath the switch

(iv)   where more than one switch is installed on any one building, each switch shall be clearly marked to indicate the installation or part of the installation which it controls.

PART 5
SELECTION AND ERECTION OF EQUIPMENT

CONTENTS

CHAPTER 51	**COMMON RULES**	
510	GENERAL	
511	COMPLIANCE WITH STANDARDS	
512	OPERATIONAL CONDITIONS AND EXTERNAL INFLUENCES	
513	ACCESSIBILITY	
514	IDENTIFICATION AND NOTICES	
515	MUTUAL DETRIMENTAL INFLUENCE	
CHAPTER 52	**SELECTION AND ERECTION OF WIRING SYSTEMS**	
521	SELECTION OF TYPE OF WIRING SYSTEM	
522	SELECTION AND ERECTION IN RELATION TO EXTERNAL INFLUENCES	
523	CURRENT-CARRYING CAPACITY OF CONDUCTORS	
524	CROSS-SECTIONAL AREAS OF CONDUCTORS	
525	VOLTAGE DROP IN CONSUMERS' INSTALLATIONS	
526	ELECTRICAL CONNECTIONS	
527	SELECTION AND ERECTION TO MINIMISE THE SPREAD OF FIRE	
528	PROXIMITY TO OTHER SERVICES	
529	SELECTION AND ERECTION IN RELATION TO MAINTAINABILITY INCLUDING CLEANING	
CHAPTER 53	**SWITCHGEAR (for protection, isolation and switching)**	
530	COMMON REQUIREMENTS	
531	DEVICES FOR PROTECTION AGAINST ELECTRIC SHOCK	
532	*(Reserved for future use)*	
533	OVERCURRENT PROTECTIVE DEVICES	
534	*(Reserved for future use)*	

535	DEVICES FOR PROTECTION AGAINST UNDERVOLTAGE	
536	*(Reserved for future use)*	
537	ISOLATING AND SWITCHING DEVICES	
CHAPTER 54	**EARTHING ARRANGEMENTS AND PROTECTIVE CONDUCTORS**	
541	GENERAL	
542	CONNECTIONS TO EARTH	
543	PROTECTIVE CONDUCTORS	
544	*(Reserved for future use)*	
545	EARTHING ARRANGEMENTS FOR FUNCTIONAL PURPOSES	
546	EARTHING ARRANGEMENTS FOR COMBINED PROTECTIVE AND FUNCTIONAL PURPOSES	
547	PROTECTIVE BONDING CONDUCTORS	
CHAPTER 55	**OTHER EQUIPMENT**	
551	TRANSFORMERS	
552	ROTATING MACHINES	
553	ACCESSORIES	
554	CURRENT-USING EQUIPMENT	
CHAPTER 56	**SUPPLIES FOR SAFETY SERVICES**	
561	GENERAL	
562	SOURCES	
563	CIRCUITS	
564	UTILISATION EQUIPMENT	
565	SPECIAL REQUIREMENTS FOR SAFETY SERVICES HAVING SOURCES NOT CAPABLE OF OPERATION IN PARALLEL	
566	SPECIAL REQUIREMENTS FOR SAFETY SERVICES HAVING SOURCES CAPABLE OF OPERATION IN PARALLEL	

CHAPTER 51

COMMON RULES

CONTENTS

510	GENERAL	
511	COMPLIANCE WITH STANDARDS	
512	OPERATIONAL CONDITIONS AND EXTERNAL INFLUENCES	
512-01	**Voltage**	
512-02	**Current**	
512-03	**Frequency**	
512-04	**Power**	
512-05	**Compatibility**	
512-06	**External influences**	
513	ACCESSIBILITY	
514	IDENTIFICATION AND NOTICES	
514-01	**General**	
514-02	**Conduit**	
514-03 to 514-07	*Identification of conductors*	
514-03	**Protective conductor**	
514-04	**Neutral conductor**	
514-05	**PEN conductor**	
514-06	**Non-flexible cable and conductors**	
514-07	**Flexible cable and flexible cord**	
514-08	**Identification of a protective device**	
514-09	**Diagrams**	
514-10	**Warning notice - voltage**	
514-11	**Warning notice - isolation**	
514-12	**Notice - periodic inspection and testing**	
514-13	**Warning notice - earthing and bonding connections**	
515	MUTUAL DETRIMENTAL INFLUENCE	

PART 5

SELECTION AND ERECTION OF EQUIPMENT

CHAPTER 51

COMMON RULES

510 GENERAL

510-01-01 Every item of equipment shall be selected and erected so as to comply with the requirements stated in the following Regulations of this Chapter and the relevant Regulations in other Chapters of the Regulations.

511 COMPLIANCE WITH STANDARDS

511-01-01 Every item of equipment shall comply with the relevant requirements of the applicable British Standard, or harmonized European Standard appropriate to the intended use of the equipment. The edition of the Standard shall be the current edition, with those amendments pertaining at a date to be agreed by the parties to the contract concerned.

Alternatively, if equipment complying with a foreign national standard based on an IEC Standard is to be used the designer or other person responsible for specifying the installation shall verify that any differences between that standard and the corresponding British Standard or harmonized European Standard will not result in a lesser degree of safety than that afforded by compliance with the British Standard.

511-01-02 Where equipment to be used is not covered by a British Standard or harmonized European Standard or is used outside the scope of its standard, the designer or other person responsible for specifying the installation shall confirm that the equipment provides the same degree of safety as that afforded by compliance with the Regulations.

512 OPERATIONAL CONDITIONS AND EXTERNAL INFLUENCES

512-01 Voltage

512-01-01 Every item of equipment shall be suitable for the nominal voltage U_o of the installation or the part of the installation concerned, where necessary taking account of the highest and/or lowest voltage likely to occur in normal service. In an IT system, equipment shall be insulated for the nominal voltage between phases.

512-02 Current

512-02-01 Every item of equipment shall be suitable for:

(i) the design current, taking into account any capacitive and inductive effects

(ii) the current likely to flow in abnormal conditions for such periods of time as are determined by the characteristics of the protective devices concerned.

Where switchgear, protective devices, accessories and other types of equipment are to be connected to conductors intended to operate at a temperature exceeding 70°C in normal service, the equipment manufacturer shall be consulted to determine if the nominal current rating of the equipment may need to be appropriately reduced.

512-03 Frequency

512-03-01 If frequency has an influence on the characteristics of the equipment, the rated frequency of the equipment shall correspond to the nominal frequency of the supply to the circuit concerned.

512-04 Power

512-04-01 Every item of equipment selected on the basis of its power characteristics shall be suitable for the duty demanded of the equipment.

512-05 Compatibility

512-05-01 Every item of equipment shall be selected and erected so that it will neither cause harmful effects to other equipment nor impair the supply during normal service including switching operations.

512-06 External influences

512-06-01 Every item of equipment shall be of a design appropriate to the situation in which it is to be used and/or its mode of installation shall take account of the conditions likely to be encountered, including tests required by Part 7.

If the equipment does not, by its construction, have the characteristics relevant to the external influences of its location, it shall be provided with appropriate additional protection in the erection of the installation. Such protection shall not adversely affect the operation of the equipment thus protected.

512-06-02 Where different external influences occur simultaneously the degree of protection provided shall take account of any mutual effect.

513 ACCESSIBILITY

513-01-01 Except for a joint in cables where Section 526 allows such a joint to be inaccessible, every item of equipment shall be arranged so as to facilitate its operation, inspection and maintenance and access to each connection. Such facilities shall not be significantly impaired by mounting equipment in an enclosure or a compartment.

514 IDENTIFICATION AND NOTICES

514-01 General

514-01-01 Except where there is no possibility of confusion, a label or other suitable means of identification shall be provided to indicate the purpose of each item of switchgear and controlgear.

Where the operation of switchgear and controlgear cannot be observed by the operator and where this might lead to danger, a suitable indicator, complying with BS 4099 where applicable, shall be fixed in a position visible to the operator.

514-02 Conduit

514-02-01 Where an electrical conduit is required to be distinguished from a pipeline or another service, orange shall be used as the basic identification colour in compliance with BS 1710.

Identification of conductors

514-03 Protective conductor

514-03-01 The colour combination green and yellow is reserved exclusively for identification of a protective conductor and shall not be used for any other purpose. In this combination one of the colours shall cover at least 30% and at most 70% of the surface being coloured, while the other colour covers the remainder of the surface.

A bare conductor or busbar used as protective conductor shall be identified, where necessary, by equal green-and-yellow stripes, each not less than 15 mm and not more than 100 mm wide, close together, either throughout the length of the conductor or in each compartment and unit and at each accessible position. If adhesive tape is used, it shall be bi- coloured.

514-04 Neutral Conductor

514-04-01 *(Reserved for future use)*

514-05 PEN conductor

514-05-01 *(Reserved for future use)*

514-06 Non-flexible cables and conductors

514-06-01 Every single-core non-flexible cable and every core of non-flexible cable for use as fixed wiring shall be identifiable at its terminations and preferably throughout its length, by the appropriate method described in Items (i) to (v) below :

(i) for rubber-insulated and p.v.c.-insulated cables, the use of core colours in accordance with the requirements of Table 51A, or the application at terminations of tapes, sleeves or discs of the appropriate colours prescribed in the Table

(ii) for armoured p.v.c.-insulated auxiliary cables, as an alternative to the method described in Item (i) above, the use of numbered cores in accordance with BS 6346

(iii) for paper-insulated cables, the use of numbered cores in accordance with BS 6480; provided that the numbers 1, 2 and 3 shall signify phase conductors, the number 0 the neutral conductor, and the number 4 the fifth ('special-purpose') core if any

(iv) for cables with thermosetting insulation, the use of core colours in accordance with the requirements of Table 51A, or alternatively the use of numbered cores in accordance with BS 5467 and BS 6724 provided that the numbers 1, 2 and 3 shall signify phase conductors, and the number 0 the neutral conductor

(v) for mineral-insulated cables, the application at terminations of tapes, sleeves or discs of the appropriate colours prescribed in Table 51A.

Binding and sleeving for identification purposes shall comply with BS 3858 where appropriate.

514-06-02 The single colour green shall not be used.

514-06-03 A bare conductor shall be identified, where necessary, by the application of tape, sleeve, or disc of the appropriate colour prescribed in Table 51A, or by painting with such a colour.

514-06-04 Any colour used to identify a switchboard busbar or conductor shall comply with the requirements of Table 51A so far as these are applicable.

TABLE 51A

Colour identification of cores of non-flexible cables and bare conductors for fixed wiring
For armoured p.v.c.-insulated cables and paper-insulated cables, see Regulation 514-06-01(ii) and (iii).

Function	Colour identification
Protective (including earthing) conductor	green-and-yellow
Phase of a.c. single-phase circuit	red (or yellow or blue*)
Neutral of a.c. single- or three-phase circuit	black
Phase R of 3-phase a.c. circuit	red
Phase Y of 3-phase a.c. circuit	yellow
Phase B of 3-phase a.c. circuit	blue
Positive of d.c. 2-wire circuit	red
Negative of d.c. 2-wire circuit	black
Outer (positive or negative) of d.c. 2-wire circuit derived from 3-wire system	red
Positive of 3-wire d.c. circuit	red
Middle wire of 3-wire d.c. circuit	black
Negative of 3-wire d.c. circuit	blue
Functional Earth - Telecommunications	cream (See BS 6701 Part 1)

* As alternatives to the use of red, if desired, in large installations, on the supply side of the final distribution board.

514-07   Flexible cable and flexible cords

514-07-01 Every core of a flexible cable or flexible cord shall be identifiable throughout its length as appropriate to its function, as indicated in Table 51B.

514-07-02 Flexible cable or flexible cord containing a core of the single colour green, yellow alone or any bi-colour other than the colour combination green-and-yellow, shall not be used (see Regulation 514-06-02).

TABLE 51B
Colour identification of cores of flexible cables and flexible cords

Number of cores	Function of core	Colour(s) of core
1	Phase Neutral Protective	Brown[1] Blue Green-and-yellow
2	Phase Neutral	Brown Blue[2]
3	Phase Neutral Protective	Brown[3] Blue[2] Green-and-yellow
4 or 5	Phase Neutral Protective	Brown or black[4] Blue[2] Green-and-yellow

(1) Or any other colour not prohibited by Regulations 514-03-01 and 514-07-02, excepting blue.

(2) The blue core may be used for functions other than the neutral in a circuit which does not incorporate a neutral conductor, in which case its function shall be appropriately identified during installation; provided that the blue core shall not in any event be used as a protective conductor. If the blue core is used for another function, the coding L1, L2, L3, or other coding where appropriate shall be used.

(3) In a three-core flexible cable or a flexible cord not incorporating a green-and-yellow core, a brown core and a black core may be used as phase conductors.

(4) Where an indication of phase rotation is desired, or it is desired to distinguish the function of more than one conductor of the same colour, this shall be by the application of a numbered or lettered (not coloured) sleeve to the core, preferably using the coding L1, L2, L3 or other coding where appropriate.

514-08 Identification of a protective device

514-08-01 A protective device shall be arranged and identified so that the circuit protected may be easily recognised.

514-09 Diagrams

514-09-01 A legible diagram, chart or table or equivalent form of information shall be provided indicating in particular:

(i) the type and composition of each circuit (points of utilisation served, number and size of conductors, type of wiring), and

(ii) a description of the method used for compliance with Regulation 413-02-04, and

(iii) the information necessary for the identification of each device performing the functions of protection, isolation and switching, and its location, and

(iv) any circuit or equipment vulnerable to a typical test.

For simple installations the foregoing information may be given in a schedule. A durable copy of the schedule relating to a distribution board shall be provided within or adjacent to each distribution board.

Any symbol used shall comply with BS 3939.

514-10 Warning notice - voltage

514-10-01 Every item of equipment or enclosure within which a voltage exceeding 250 volts exists, and where the presence of such a voltage would not normally be expected, shall be so arranged that before access is gained to a live part, a warning of the maximum voltage present is clearly visible.

Where terminals or other fixed live parts between which a voltage exceeding 250 volts exists are housed in separate enclosures or items of equipment which, although separated, can be reached simultaneously by a person, a notice shall be secured in a position that anyone, before gaining access to such live parts, is warned of the maximum voltage which exists between those parts.

Means of access to all live parts of switchgear and other fixed live parts where different nominal voltages exist shall be marked to indicate the voltages present.

514-11 Warning notice - isolation

514-11-01 A notice of durable material in accordance with Regulation 461-01-05, shall be fixed in each position where there are live parts which are not capable of being isolated by a single device. The location of each isolator (disconnector) shall be indicated unless there is no possibility of confusion.

514-12 Notice - periodic inspection and testing

514-12-01 A notice of such durable material as to be likely to remain easily legible throughout the life of the installation, shall be fixed in a prominent position at or near the origin of every installation upon completion of the work carried out in accordance with Chapters 71 or 73.

The notice shall be inscribed in indelible characters not smaller than those here illustrated and shall read as follows:

IMPORTANT

This installation should be periodically inspected and tested and a report on its condition obtained, as prescribed in the Regulations for Electrical Installations issued by the Institution of Electrical Engineers.

Date of last inspection......................................

Recommended date of next inspection..........................

514-12-02 Where an installation incorporates a residual current device a notice shall be fixed in a prominent position at or near the origin of the installation. The notice shall be in indelible characters not smaller than those here illustrated and shall read as follows:

This installation, or part of it, is protected by a device which automatically switches off the supply if an earth fault develops. Test quarterly by pressing the button marked "T" or "Test". The device should switch off the supply and should then be switched on to restore the supply. If the device does not switch off the supply when the button is pressed, seek expert advice.

514-13   Warning notice - earthing and bonding connections

514-13-01 A permanent label durably marked with the words as follows and no smaller than the example shall be permanently fixed in a visible position at or near :-

(i)   the point of connection of every earthing conductor to an earth electrode, and

(ii) the point of connection of every bonding conductor to an extraneous-conductive-part.

Safety Electrical Connection - Do Not Remove

514-13-02 Where Regulations 471-11 or 471-12 apply, the warning notice specified shall be durably marked in legible type not smaller than that illustrated here and shall read as follows:

The equipotential protective bonding conductors associated with the electrical installation in this location MUST NOT BE CONNECTED TO EARTH.

Equipment having exposed-conductive-parts connected to earth must not be brought into this location.

---

515   MUTUAL DETRIMENTAL INFLUENCE

515-01-01 Electrical equipment shall be selected and erected so as to avoid any harmful influence between the electrical installation and any non-electrical installations envisaged.

515-01-02 Equipment between which there may be mutual detrimental influence shall, where necessary, be effectively segregated.

CHAPTER 52

SELECTION AND ERECTION OF WIRING SYSTEMS

CONTENTS

521	SELECTION OF TYPE OF WIRING SYSTEM
521-01	**Cables and conductors for low voltage**
521-02	**Cables for a.c. circuits - electromagnetic effects**
521-03	**Electromechanical stresses**
521-04	**Conduits and conduit fittings**
521-05	**Trunking, ducting and fittings**
521-06	**Lighting track systems**
521-07	**Methods of installation of cables and conductors**
522	SELECTION AND ERECTION IN RELATION TO EXTERNAL INFLUENCES
522-01	**Ambient temperature (AA)**
522-02	**External heat sources**
522-03	**Presence of water (AD) or high humidity (AB)**
522-04	**Presence of solid foreign bodies (AE)**
522-05	**Presence of corrosive or polluting substances (AF)**
522-06	**Impact (AG)**
522-07	**Vibration (AH)**
522-08	**Other mechanical stresses (AJ)**
522-09	**Presence of flora and/or mould growth (AK)**
522-10	**Presence of fauna (AL)**
522-11	**Solar radiation (AN)**
522-12	**Building design (CB)**
523	CURRENT-CARRYING CAPACITY OF CONDUCTORS
523-01	**Conductor operating temperature**
523-02	**Cables connected in parallel**
523-03	**Cables connected to bare conductors or busbars**
523-04	**Cables in thermal insulation**
523-05	**Metallic sheaths and/or non-magnetic armour of single-core cables**

524	CROSS-SECTIONAL AREAS OF CONDUCTORS
524-01	**Phase conductors in a.c. circuits and live conductors in d.c. circuits**
524-02	**Neutral conductors**
525	VOLTAGE DROP IN CONSUMERS' INSTALLATIONS
525-01	**Voltage drop in consumers' installations**
526	ELECTRICAL CONNECTIONS
526-01	**Connections between conductors and between a conductor and equipment**
526-02	**Selection of means of connection**
526-03	**Enclosed connections**
526-04	**Accessibility of connections**
527	SELECTION AND ERECTION TO MINIMISE THE SPREAD OF FIRE
527-01	**Risk of spread of fire**
527-02	**Sealing of wiring system penetrations**
527-03	**Erection conditions**
527-04	**Verification**
528	PROXIMITY TO OTHER SERVICES
528-01	**Proximity to electrical services**
528-02	**Proximity to non-electrical services**
529	SELECTION AND ERECTION IN RELATION TO MAINTAINABILITY, INCLUDING CLEANING

CHAPTER 52

SELECTION AND ERECTION OF WIRING SYSTEMS

521 SELECTION OF TYPE OF WIRING SYSTEM

521-01 Cables and conductors for low voltage

521-01-01 Every non-flexible or flexible cable or flexible cord for use at low voltage shall comply with the appropriate British Standard (see Regulation 120-04).

For aerial use or suspension any non-flexible cable sheathed with lead, p.v.c. or an elastomeric material may incorporate a catenary wire or include hard-drawn copper conductors.

This Regulation does not apply to a flexible cord forming part of a portable appliance or luminaire where the appliance or luminaire as a whole is the subject of and complies with a British Standard, or to special flexible cables and flexible cords for combined power and telecommunication wiring (see Regulation 528-01-08).

Insulated flexible cable and flexible cord may incorporate a flexible metallic armour, braid or screen.

521-01-02 A busbar trunking system shall comply with BS 5486, Part 2.

521-01-03 Every conductor, other than a cable, for use as an overhead line operating at low voltage shall comply with an appropriate British Standard for overhead conductors.

521-01-04 A flexible cable shall be used for fixed wiring only where the relevant provisions of the Regulations are met.

521-02 Cables for a.c. circuits - electromagnetic effects

521-02-01 Single-core cables armoured with steel wire or tape shall not be used for a.c. circuits. Conductors of a.c. circuits installed in ferromagnetic enclosures shall be arranged so that the conductors of all phases and the neutral conductor (if any) and the appropriate protective conductor of each circuit are contained in the same enclosure.

Where such conductors enter a ferrous enclosure they shall be arranged so that the conductors are not individually surrounded by a ferrous material, or other provision shall be made to prevent eddy (induced) currents.

521-03 Electromechanical stresses

521-03-01 Every conductor or cable shall have adequate strength and be so installed as to withstand the electromechanical forces that may be caused by any current, including fault current, it may have to carry in service.

521-04 Conduits and conduit fittings

521-04-01 A conduit or conduit fitting shall comply with the appropriate British Standard referred to below:

  (i) steel conduit and fittings - BS 31, BS 6053, BS 6099 Part 1

  (ii) flexible steel conduit - BS 731 Part 1, BS 6053, BS 6099 Part 1

  (iii) steel conduit fittings with metric threads - BS 4568, BS 6053, BS 6099 Part 1

  (iv) non-metallic conduits and fittings - BS 4607, BS 6053, BS 6099 Part 2, Section 2.2.

521-05 Trunking, ducting and fittings

521-05-01 Where applicable, trunking, ducting and their fittings shall comply with BS 4678. Where BS 4678 does not apply, non-metallic trunking, ducting and their fittings shall be of insulating material complying with the ignitability characteristic 'P' of BS 476 Part 5.

521-06 Lighting track systems

521-06-01 A lighting track system shall comply with BS 4533 Part 2, Section 2.6.

521-07 Methods of installation of cables and conductors

521-07-01 The methods of installation of a wiring system for which the Regulations specifically provide are shown in Appendix 4 Table 4A.

The use of another method is not precluded provided that compliance with the Regulations is maintained.

521-07-02 A bare live conductor shall be installed on insulators.

521-07-03 Non-sheathed cables for fixed wiring shall be enclosed in conduit, ducting or trunking. This Regulation does not apply to a protective conductor complying with Section 543 of the Regulations.

Where cables having different temperature ratings are installed in the same enclosure, all the cables shall be deemed to have the lowest temperature rating.

522 SELECTION AND ERECTION IN RELATION TO EXTERNAL INFLUENCES
(See Chapter 32 and Appendix 5).

522-01 Ambient temperature (AA)

522-01-01 A wiring system shall be selected and erected so as to be suitable for the highest and lowest local ambient temperature likely to be encountered.

522-01-02 The components of a wiring system, including cables and wiring enclosures shall be installed or handled only at temperatures within the limits stated in the relevant product specification or as recommended by the manufacturer.

522-02 External heat sources

522-02-01 To avoid the effects of heat from external sources including solar gain one or more of the following methods, or an equally effective method, shall be used to protect the wiring system:

   (i) shielding

   (ii) placing sufficiently far from the source of heat

   (iii) selecting a system with due regard for the additional temperature rise which may occur

   (iv) reducing the current-carrying capacity

   (v) local reinforcement or substitution of insulating material.

522-02-02 Parts of a cable or flexible cord within an accessory, appliance or luminaire shall be suitable for the temperatures likely to be encountered, as determined in accordance with Regulation 522-01-01, or shall be provided with additional insulation suitable for those temperatures.

522-03 Presence of Water (AD) or High Humidity (AB)

522-03-01 A wiring system shall be selected and erected so that no damage is caused by high humidity or ingress of water during installation, use and maintenance.

522-03-02 Where water may collect or condensation may form in a wiring system provision shall be made for its harmless escape through suitably located drainage points.

522-03-03 Where a wiring system may be subjected to waves (AD6), protection against mechanical damage shall be afforded by one or more of the methods given in Regulations 522-06 to 522-08.

522-04 Presence of solid foreign bodies (AE)

522-04-01 A wiring system shall be selected and erected to minimise the ingress of solid foreign bodies during installation, use and maintenance.

522-04-02 In a location where dust or other substance in significant quantity may be present (AE4) additional precautions shall be taken to prevent its accumulation in quantities which could adversely affect the heat dissipation from the wiring system.

522-05 Presence of corrosive or polluting substances (AF)

522-05-01 Where the presence of corrosive or polluting substances is likely to give rise to corrosion or deterioration, parts of the wiring system likely to be affected shall be suitably protected or manufactured from materials resistant to such substances.

522-05-02 Metals liable to initiate electrolytic action shall not be placed in contact with each other.

522-05-03 Materials liable to cause mutual or individual deterioration or hazardous degradation shall not be placed in contact with each other.

522-06 Impact (AG)

522-06-01 A wiring system shall be selected and erected so as to minimise mechanical damage.

522-06-02 In a fixed installation where an impact of medium severity (AG2) or high severity (AG3) can occur, protection shall be afforded by:

(i) the mechanical characteristics of the wiring system, or

(ii) the location selected, or

(iii) the provision of additional local or general mechanical protection,

or by any combination of the above.

522-06-03 Except where installed in a conduit or duct which provides equivalent mechanical protection, a cable buried in the ground shall be of a construction incorporating an armour or metal sheath or both, or be of insulated concentric construction. Such cable shall be marked by cable covers or a suitable marking tape or by suitable identification of the conduit or duct and be buried at a sufficient depth to avoid being damaged by any disturbance of the ground reasonably likely to occur.

522-06-04 A wiring system buried in a floor shall be sufficiently protected to prevent damage caused by the intended use of the floor.

522-06-05 Where a cable is installed under a floor or above a ceiling it shall be run in such a position that it is not liable to be damaged by contact with the floor or the ceiling or their fixings. Where a cable passes through a timber joist within a floor or ceiling construction or through a ceiling support (e.g. under floorboards), the cable shall be at least 50 mm measured vertically from the top, or bottom as appropriate, of the joist or batten. Alternatively, cable shall incorporate an earthed metallic sheath suitable for use as a protective conductor or shall be protected by enclosure in earthed steel conduit securely supported, or by equivalent mechanical protection sufficient to prevent penetration of the cable by nails, screws, and the like (see also Regulation 471-13-04(v)).

522-06-06 Where a cable is to be concealed within a wall or partition at a depth of less than 50 mm from the surface its method of erection shall be as follows or its construction shall comply with Regulation 522-06-07 as appropriate. The cable shall be installed within 150 mm of the top of the wall or partition or within 150 mm of an angle formed by two adjoining walls or partitions. Where the cable is connected to a point or accessory on the wall or partition, the cable may be installed outside these zones only in a straight run, either horizontally or vertically to the point or accessory.

522-06-07 Where compliance with Regulation 522-06-06 is impracticable, the concealed cable shall incorporate an earthed metallic covering which complies with the requirements of these Regulations for a protective conductor of the circuit concerned, or shall be enclosed in earthed conduit, trunking or ducting satisfying the requirements of these Regulations for a protective conductor, or by mechanical protection sufficient to prevent penetration of the cable by nails, screws and the like.

522-07 Vibration (AH)

522-07-01 A wiring system supported by, or fixed to, a structure or equipment subject to vibration of medium severity (AH2) or high severity (AH3) shall be suitable for the conditions and in particular shall employ cable with fixings and connections suitable for such a situation.

522-08 Other mechanical stresses (AJ)

522-08-01 A wiring system shall be selected and erected so as to minimise during installation, use and maintenance, damage to the sheath and insulation of cables and insulated conductors and their terminations.

522-08-02 There shall be adequate means of access for drawing cable in or out and, if buried in the structure, a conduit or cable ducting system for each circuit shall be completely erected before cable is drawn in.

522-08-03 The radius of every bend in a wiring system shall be such that conductors and cables shall not suffer damage.

522-08-04 Where a conductor or a cable is not continuously supported it shall be supported by suitable means at appropriate intervals in such a manner that the conductor or cable does not suffer damage by its own weight.

522-08-05 Every cable or conductor used as fixed wiring shall be supported in such a way that it is not exposed to undue mechanical strain and so that there is no appreciable mechanical strain on the terminations of the conductors, account being taken of mechanical strain imposed by the supported weight of the cable or conductor itself.

522-08-06 A flexible wiring system shall be installed so that excessive tensile and torsional stresses to the conductors and connections are avoided.

522-09 Presence of flora and/or mould growth (AK)

522-09-01 Where expected conditions constitute a hazard (AK2), the wiring system shall be selected accordingly or special protective measures shall be adopted.

522-10 Presence of fauna (AL)

522-10-01 Where expected conditions constitute a hazard (AL2), the wiring system shall be selected accordingly or special protective measures shall be adopted.

522-11 Solar radiation (AN)

522-11-01 Where significant solar radiation (AN2) is experienced or expected, a wiring system suitable for the conditions shall be selected and erected or adequate shielding shall be provided.

522-12 Building design (CB)

522-12-01 Where structural movement (CB3) is experienced or expected, the cable support and protection system employed shall be capable of permitting relative movement so that conductors are not subjected to excessive mechanical stress.

522-12-02 For flexible or unstable structures (CB4) flexible wiring systems shall be used.

523 CURRENT-CARRYING CAPACITY OF CONDUCTORS

523-01 Conductor operating temperature

523-01-01 The current to be carried by any conductor for sustained periods during normal operation shall be such that the conductor operating temperature given in the appropriate table of current-carrying capacity in Appendix 4 is not exceeded (see Table 52B).

Where a conductor operates at a temperature exceeding 70°C it shall be ascertained that the equipment connected to the conductor is suitable for the conductor operating temperature (see Regulation 512-02).

523-02 Cables connected in parallel

523-02-01 Except for a ring final circuit, cables connected in parallel shall be of the same construction, cross-sectional area, length and disposition, without branch circuits and arranged so as to carry substantially equal currents.

523-03 Cables connected to bare conductors or busbars

523-03-01 Where a cable is to be connected to a bare conductor or busbar its type of insulation and/or sheath shall be suitable for the maximum operating temperature of the bare conductor or busbar.

523-04 Cables in thermal insulation

523-04-01 Where a cable is to be run in a space to which thermal insulation is likely to be applied, the cable shall wherever practicable be fixed in a position such that it will not be covered by the thermal insulation. Where fixing in such a position is impracticable the cross-sectional area of the cable shall be appropriately increased.

For a cable installed in a thermally insulated wall or above a thermally insulated ceiling, the cable being in contact with a thermally conductive surface on one side, current-carrying capacities are tabulated in Appendix 4, Method 4 being the appropriate Reference Method.

For a single cable likely to be totally surrounded by thermally insulating material over a length of more than 0.5 m, the current-carrying capacity shall be taken, in the absence of more precise information, as 0.5 times the current-carrying capacity for that cable clipped direct to a surface and open (Reference Method 1).

Where a cable is to be totally surrounded by thermal insulation for less than 0.5 m the current-carrying capacity of the cable shall be reduced appropriately depending on the size of cable, length in insulation and thermal properties of the insulation. The derating factors in Table 52A are appropriate to conductor sizes up to 10 mm^2 in thermal insulation having a thermal conductivity greater than 0.0625 W/K.m.

TABLE 52A

Cable surrounded by thermal insulation

length in insulation mm	derating factor
50	0.89
100	0.81
200	0.68
400	0.55

TABLE 52B (Regulation 523-01)

Maximum conductor operating temperatures

Conductor material	Insulation material	Conductor operating temperature °C	Limiting final temperature °C	Appendix 4 Table
Copper	70 °C p.v.c. (General purpose) 60 °C rubber 85 °C rubber 85 °C p.v.c. 90 °C thermosetting Impregnated paper	70 60 85 85 90 80	160/140* 200 220 160/140* 250 160	4D 1-2-3 & 4 4H1 4F 1 & 2 4E 1-2-3 & 4
Copper	Mineral - plastic covered or exposed to touch - bare and neither exposed to touch nor in contact with combustible materials	70 (sheath) 105 (sheath)	160 250	4J1 4J2
Aluminium	70 °C p.v.c. (General purpose) 85 °C p.v.c. 60 °C rubber 85 °C rubber 90 °C thermosetting Impregnated paper	70 85 60 85 90 80	160/140* 160/140* 200 220 250 160	4K 1-2-3 & 4 4L1-2-3 & 4

* above 300 mm^2

TABLE 52C (Regulation 524-01)

MINIMUM NOMINAL CROSS-SECTIONAL AREA OF CONDUCTOR

TYPE OF WIRING SYSTEM	USE OF THE CIRCUIT	CONDUCTOR	
		MATERIAL	MINIMUM PERMISSIBLE NOMINAL CROSS-SECTIONAL AREA mm^2
Cables and insulated conductors	Power and lighting circuits	Cu Al	1.0 16.0 (see Note 1)
	Signalling and control circuits	Cu	0.5 (see Note 2)
Bare conductors	Power circuits	Cu Al	10 16
	Signalling and control circuits	Cu	4
Flexible connections with insulated conductors and cables	For a specific appliance		As specified in the relevant British Standard
	For any other application	Cu	0.5 (see Note 2)
	Extra-low voltage circuits for special applications		0.5

NOTES:
1. Connectors used to terminate aluminium conductors shall be tested and approved for this specific use.
2. In multicore flexible cables containing 7 or more cores and in signalling control circuits intended for electronic equipment a minimum nominal cross-sectional area of 0.1 mm^2 is permitted

523-05 Metallic sheaths and/or non-magnetic armour of single-core cables

523-05-01 The metallic sheaths and/or non-magnetic armour of single-core cables in the same circuit shall normally be bonded together at both ends of their run (solid bonding). Alternatively the sheaths or armour of such cables having conductors of cross-sectional area exceeding 50 mm^2 and a non-conducting outer sheath may be bonded together at one point in their run (single point bonding) with suitable insulation at the un-bonded ends, in which case the length of the cables from the bonding point shall be limited so that, at full load, voltages from sheaths and/or armour to Earth:

(i) do not exceed 25 volts,

(ii) do not cause corrosion when the cables are carrying their full load current, and

(iii) do not cause danger or damage to property when the cables are carrying short-circuit current.

524 CROSS-SECTIONAL AREAS OF CONDUCTORS

524-01 Phase conductors in a.c. circuits and live conductors in d.c.circuits

524-01-01 The nominal cross-sectional area of phase conductors in a.c. circuits and of live conductors in d.c. circuits shall be not less than the values specified in Table 52C.

524-02 Neutral conductors

524-02-01 For a polyphase circuit in which imbalance may occur in normal service, through significant inequality of loading or of power factor in the various phases, or through the presence of significant harmonic currents in the various phases, the neutral conductor shall have a cross-sectional area adequate to afford compliance with Regulation 523-01 for the maximum current likely to flow in it (see Section 473 for overcurrent protection of neutral conductors).

524-02-02 For a polyphase circuit in which serious imbalance is unlikely to occur in normal service, other than a discharge lighting circuit, multicore cables incorporating a reduced neutral conductor in accordance with the appropriate British Standard may be used. Where single-core cables are used in such circuits, the neutral conductor shall have a cross-sectional area appropriate to the expected value of the neutral current.

524-02-03 In a discharge lighting circuit the neutral conductor shall have a cross-sectional area not less than that of the phase conductor(s).

525 VOLTAGE DROP IN CONSUMERS' INSTALLATIONS

525-01 Voltage drop in consumers' installations

525-01-01 Under normal service conditions the voltage at the terminals of any fixed current-using equipment shall be greater than the lower limit corresponding to the British Standard relevant to the equipment.

Where the fixed current-using equipment concerned is not the subject of a British Standard the voltage at the terminals shall be such as not to impair the safe functioning of that equipment.

525-01-02 The requirements of Regulation 525-01-01 are deemed to be satisfied for a supply given in accordance with the Electricity Supply Regulations 1988 (as amended) if the voltage drop between the origin of the installation (usually the supply terminals) and the fixed current-using equipment does not exceed 4% of the nominal voltage of the supply.

A greater voltage drop may be accepted for a motor during starting periods and for other equipment with high inrush currents provided that it is verified that the voltage variations are within the limits specified in the relevant British Standards for the equipment or, in the absence of a British Standard, in accordance with the manufacturer's recommendations.

526 ELECTRICAL CONNECTIONS

526-01 Connections between conductors and between a conductor and equipment

526-01-01 Every connection between conductors and between a conductor and equipment shall provide durable electrical continuity and adequate mechanical strength (see other Mechanical Stresses, Regulation 522-08).

526-02 Selection of means of connection

526-02-01 The selection of the means of connection shall take account, as appropriate, of the following:

(i) the material of the conductor and its insulation

(ii) the number and shape of the wires forming the conductor

(iii) the cross-sectional area of the conductor

(iv) the number of conductors to be connected together

(v) the temperature attained by the terminals in normal service such that the effectiveness of the insulation of the conductors connected to them is not impaired

(vi) where a soldered connection is used the design shall take account of creep, mechanical stress and temperature rise under fault current conditions

(vii) the provision of adequate locking arrangements in situations subject to vibration or thermal cycling.

526-03 Enclosed connections

526-03-01 Where a connection is made in an enclosure the enclosure shall provide adequate mechanical protection and protection against relevant external influences.

526-03-02 Every termination and joint in a live conductor or a PEN conductor shall be made within one of the following or a combination thereof:

(i) a suitable accessory complying with the appropriate British Standard

(ii) an equipment enclosure, complying with the appropriate British Standard

(iii) a suitable enclosure of material complying with the relevant glow wire test requirements of BS 6458, Section 2.1

(iv) an enclosure formed or completed with building material considered to be non-combustible when tested to BS 476 Part 4

(v) an enclosure formed or completed by part of the building structure, having the ignitability characteristic 'P' as specified in BS 476 Part 5.

526-03-03 Cores of sheathed cables from which the sheath has been removed and non-sheathed cables at the termination of conduit, ducting or trunking shall be enclosed as required by Regulation 526-03-02.

526-04 Accessibility of connections

526-04-01 Except for the following, every connection and joint shall be accessible for inspection, testing and maintenance:

(i) a compound-filled or encapsulated joint

(ii) a connection between a cold tail and a heating element (e.g. a ceiling and floor heating system, a pipe trace-heating system)

(iii) a joint made by welding, soldering, brazing or compression tool.

527 SELECTION AND ERECTION TO MINIMISE THE SPREAD OF FIRE

527-01 Risk of spread of fire

527-01-01 The risk of spread of fire shall be minimised by selection of an appropriate material and erection in accordance with Section 527.

527-01-02 The wiring system shall be installed so that the general building structural performance and fire safety are not materially reduced.

527-01-03 A part of a wiring system which complies with the requirements of the relevant British Standard, which Standard has no requirement for testing for resistance to the propagation of flame, shall be completely enclosed in non-combustible building material having the ignitability characteristic 'P' as specified in BS 476 Part 5.

527-02 Sealing of the wiring system penetrations

527-02-01 Where a wiring system passes through elements of building construction such as floors, walls, roofs, ceilings, partitions or cavity barriers, the openings remaining after passage of the wiring system shall be sealed according to the degree of fire resistance required of the element concerned (if any).

527-02-02 Where a wiring system such as conduit, cable ducting, cable trunking, busbar or busbar trunking penetrates elements of building construction having specified fire resistance it shall be internally sealed so as to maintain the degree of fire resistance of the respective element as well as being externally sealed to maintain the required fire resistance. A non flame propagating wiring system having a maximum internal cross-section of 710 mm^2 need not be internally sealed.

Except for fire resistance over one hour, this Regulation is satisfied if the sealing of the wiring system concerned has been type tested by the method specified in BS 476 Part 23.

527-02-03 Each sealing arrangement used in accordance with Regulations 527-02-01 and 527-02-02 shall comply with the following requirements and Regulation 527-04:

(i) it shall be compatible with the material of the wiring system with which it is in contact

(ii) it shall permit thermal movement of the wiring system without reduction of the sealing quality

(iii) it shall be removable without damage to existing cable where space permits future extension to be made

(iv) it shall resist relevant external influences to the same degree as the wiring system with which it is used.

527-03 Erection conditions

527-03-01 During the erection of a wiring system temporary sealing arrangements shall be provided as appropriate.

527-03-02 During alteration work sealing which has been disturbed shall be reinstated as soon as practicable.

527-04 Verification

527-04-01 Each sealing arrangement shall be visually inspected at an appropriate time during erection to verify that it conforms to the manufacturer's erection instructions and the details shall be recorded.

528 PROXIMITY TO OTHER SERVICES

528-01 Proximity to electrical services

528-01-01 Neither an extra-low voltage nor a low voltage circuit shall be contained within the same wiring system as a circuit of nominal voltage exceeding that of low voltage unless every cable is insulated for the highest voltage present or one of the following methods is adopted:

(i) each conductor in a multicore cable is insulated for the highest voltage present in the cable, or is enclosed within an earthed metallic screen of current-carrying capacity equivalent to that of the largest conductor enclosed within the screen, or

(ii) the cables are insulated for their respective system voltage and installed in a separate compartment of a cable ducting or cable trunking system, or have an earthed metallic covering.

528-01-02 A low voltage circuit shall be separated from an extra-low voltage circuit as required by Chapter 41.

528-01-03 Where an installation comprises circuits for telecommunication, fire-alarm or emergency lighting systems, as well as circuits operating at low voltage and connected directly to a mains supply system, precautions shall be taken, in accordance with Regulations 528-01-04 to 528-01-08 and the recommendations of BS 6701, to prevent electrical contact between the cables of the various types of circuit.

528-01-04 Fire alarm and emergency lighting circuits shall be segregated from all other cables and from each other in accordance with BS 5839 and BS 5266. Telecommunication circuits shall be segregated in accordance with BS 6701 as appropriate.

528-01-05 Where a common conduit, trunking, duct or ducting is used to contain cables of category 1 and category 2 circuits, all cables of category 1 circuits shall be effectively partitioned from the cables of category 2 circuits, or alternatively the latter cables shall be insulated in accordance with the requirements of the Regulations for the highest voltage present in the category 1 circuits (see also Regulation 528-01-07).

528-01-06 Where a category 3 circuit is installed in a channel or trunking containing a circuit of any other category, the circuits shall be segregated by a continuous partition such that the specified integrity of the category 3 circuit is not reduced. Partitions shall also be provided at any common outlets in a trunking system accommodating a category 3 circuit and a circuit of another category. Where mineral-insulated cable, or cable whose performance complies with BS 6387, is used for the category 3 circuit such a partition is not normally required.

528-01-07 In conduit, duct, ducting or trunking systems, where controls or outlets for category 1 and category 2 circuits are mounted in or on a common box, switchplate or block, the cables and connections of the two categories of circuit shall be segregated by a partition which, if of metal, shall be earthed.

528-01-08 Where cores of a category 1 and a category 2 circuit are contained in a common multicore cable, flexible cable or flexible cord, the cores of the category 2 circuit shall be insulated individually or collectively as a group, in accordance with the requirements of these Regulations, for the highest voltage present in the category 1 circuit, or alternatively shall be separated from the cores of the category 1 circuit by an earthed metal screen of equivalent current-carrying capacity to that of the cores of the category 1 circuit. Where terminations of the two categories of circuit are mounted in or on a common box, switchplate, or block, they shall be segregated in accordance with Regulation 528-01-07.

528-02 Proximity to non-electrical services

528-02-01 Where a wiring system is located in close proximity to a non-electrical service both the following conditions shall be met:

(i) the wiring system shall be suitably protected against the hazards likely to arise from the presence of the other service in normal use, and

(ii) protection against indirect contact shall be afforded in accordance with the requirements of Section 413.

528-02-02 A wiring system shall not be installed in the vicinity of a service which produces heat, smoke or fume likely to be detrimental to the wiring, unless protected from harmful effects by shielding arranged so as not to affect the dissipation of heat from the wiring.

528-02-03 Where a wiring system is routed near a service liable to cause condensation (such as water, steam or gas services) precautions shall be taken to protect the wiring system from deleterious effects.

528-02-04 Where a wiring system is to be installed in proximity to a non-electrical service it shall be so arranged that any foreseeable operation carried out on either service will not cause damage to the other.

528-02-05 Any metal sheath or armour of a cable operating at low voltage, or metal conduit, duct, ducting and trunking or bare protective conductor associated with the cable which might make contact with fixed metalwork of other services shall be either segregated from it, or bonded to it.

528-02-06 No cable shall be run in a lift (or hoist) shaft unless it forms part of the lift installation as defined in BS 5655.

529 SELECTION AND ERECTION IN RELATION TO MAINTAINABILITY, INCLUDING CLEANING

529-01-01 Where any protective measure must be removed in order to carry out maintenance, reinstatement of the protective measure shall be practicable without reducing the original degree of protection.

529-01-02 Provision shall be made for safe and adequate access to all parts of the wiring system which may require maintenance.

CHAPTER 53

SWITCHGEAR
(For protection, isolation, and switching)

CONTENTS

530	COMMON REQUIREMENTS
531	DEVICES FOR PROTECTION AGAINST ELECTRIC SHOCK
531-01	**Overcurrent protective devices**
531-02	**Residual current devices**
531-03	**Residual current devices in a TN system**
531-04	**Residual current devices in a TT system**
531-05	**Residual current devices in an IT system**
531-06	**Insulation monitoring devices**
532	*(Reserved for future use)*
533	OVERCURRENT PROTECTIVE DEVICES
533-01	**Overcurrent protective devices**
533-02	**Selection of a device for the protection of a wiring system against overload**
533-03	**Selection of a device for the protection of a wiring system against fault current**
534	*(Reserved for future use)*
535	DEVICES FOR PROTECTION AGAINST UNDERVOLTAGE
536	*(Reserved for future use)*
537	ISOLATING AND SWITCHING DEVICES
537-01	**General**
537-02	**Devices for isolation**
537-03	**Devices for switching off for mechanical maintenance**
537-04	**Devices for emergency switching**
537-05	**Devices for functional switching**

CHAPTER 53

SWITCHGEAR
(For protection, isolation and switching)

530 COMMON REQUIREMENTS

530-01-01 Where an item of switchgear is required by the Regulations to disconnect all live conductors of a circuit, it shall be of a type such that the neutral conductor cannot be disconnected before the phase conductors and is reconnected before, or at the same time as, the phase conductors.

530-01-02 No fuse or, excepting where linked, switch or circuit-breaker shall be inserted in the neutral conductor of TN or TT systems.

530-01-03 A device embodying more than one function shall comply with all the requirements of this Chapter appropriate to each separate function.

531 DEVICES FOR PROTECTION AGAINST ELECTRIC SHOCK

531-01 Overcurrent protective devices

531-01-01 For a TN or a TT system, every overcurrent protective device which is to be used also for protection against electric shock (indirect contact) shall be selected so that its operating time is:

(i) appropriate to the value of fault current that would flow in the event of a fault of negligible impedance between a phase conductor and exposed-conductive-parts, so that the permissible final temperature of the phase conductor and the associated protective conductor is not exceeded (see also Regulation 543-01-01), and

(ii) appropriate for compliance with the requirements of Regulation 413-02-04.

531-01-02 For an IT system, where exposed-conductive-parts are connected together and an overcurrent protective device is to be used to provide protection against electric shock in the event of a second fault, the requirements for the protective device are the same as those for a TN system, as specified in Regulation 531-01-01.

531-02 Residual current devices

531-02-01 A residual current device shall be capable of disconnecting all the phase conductors of the circuit at substantially the same time.

531-02-02 The magnetic circuit of the transformer of a residual current device shall enclose all the live conductors of the protected circuit. The associated protective conductor shall be outside the magnetic circuit.

531-02-03 The residual operating current of the protective device shall comply with the requirements of Section 413 as appropriate to the type of system earthing.

531-02-04 A residual current device shall be so selected and the electrical circuits so subdivided that any earth leakage current which may be expected to occur during normal operation of the connected load(s) will be unlikely to cause unnecessary tripping of the device.

531-02-05 The use of a residual current device associated with a circuit normally expected to have a protective conductor, shall not be considered sufficient for protection against indirect contact if there is no such conductor, even if the rated residual operating current of the device does not exceed 30 mA.

531-02-06 A residual current device which is powered from an auxiliary source and which does not operate automatically in the case of failure of the auxiliary source shall be used only if one of the two following conditions is fulfilled:

(i) protection against indirect contact is maintained even in the case of failure of the auxiliary source, or

(ii) the device is incorporated in an installation intended to be supervised, tested and inspected by an instructed person or a skilled person.

531-02-07 A residual current device shall be located so that its operation will not be impaired by magnetic fields caused by other equipment.

531-02-08 Where a residual current device for protection against indirect contact is used with, but separately from, an overcurrent protective device, it shall be verified that the residual current operated device is capable of withstanding, without damage, the thermal and mechanical stresses to which it is likely to be subjected in the case of a fault occurring on the load side of the point at which it is installed.

531-02-09 Where, for compliance with the requirements of the Regulations for protection against indirect contact or otherwise to prevent danger, two or more residual current devices are in series, and where discrimination in their operation is necessary to prevent danger, the characteristics of the devices shall be such that the intended discrimination is achieved.

531-03   Residual current devices in a TN system

531-03-01 In a TN system, where, for certain equipment in a certain part of the installation, one or more of the conditions in Regulation 413-02-08 cannot be satisfied, that part may be protected by a residual current device.

The exposed-conductive-parts of that part of the installation shall be connected to the TN earthing system protective conductor or to a separate earth electrode which affords an impedance appropriate to the operating current of the residual current device.

In this latter case the circuit shall be treated as a TT system and Regulation 413-02-17 applies.

531-04   Residual current devices in a TT system

531-04-01 If an installation which is part of a TT system is protected by a single residual current device, this shall be placed at the origin of the installation unless the part of the installation between the origin and the device complies with the requirements for protection by the use of the Class II equipment or equivalent insulation (Regulations 413-03 and 471-09). Where there is more than one origin this requirement applies to each origin.

531-05   Residual current devices in an IT system

531-05-01 Where protection is provided by a residual current device and disconnection following a first fault is not envisaged, the non-operating residual current of the device shall be at least equal to the current which circulates on the first fault to earth of negligible impedance affecting a phase conductor.

531-06   Insulation monitoring devices

531-06-01 An insulation monitoring device shall be so designed or installed that it shall be possible to modify the setting only by the use of a key or a tool.

532   *(Reserved for future use)*

533   OVERCURRENT PROTECTIVE DEVICES

533-01   Overcurrent protective devices

533-01-01 For every fuse and circuit-breaker there shall be provided on or adjacent to it an indication of its intended nominal current as appropriate to the circuit it protects. For a semi-enclosed fuse the intended nominal current to be indicated is the value to be selected in accordance with Regulation 533-01-04.

A fuse having a fuse link likely to be replaced by a person other than a skilled person or an instructed person shall preferably be of a type such that it cannot be replaced inadvertently by one having a higher nominal current.

533-01-02 A fuse link likely to be replaced by a person other than a skilled person or an instructed person shall either:

(i)   have marked on or adjacent to it an indication of the type of fuse link intended to be used, or

(ii)   be of a type such that there is no possibility of inadvertent replacement by a fuse link having the intended nominal current but a higher fusing factor than that intended.

533-01-03 A fuse having a fuse link which is likely to be removed or replaced whilst the supply is connected shall be of a type such that it can be removed or replaced without danger.

533-01-04 A fuse shall preferably be of the cartridge type. Where a semi-enclosed fuse is selected, it shall be fitted with an element in accordance with the manufacturer's instructions if any. In the absence of such instructions, it shall be fitted with a single element of tinned copper wire of the appropriate diameter specified in Table 53A.

TABLE 53A
Sizes of tinned copper wire for use in semi-enclosed fuses.

Nominal current of fuse element (A)	Nominal diameter of wire (mm)
3	0.15
5	0.2
10	0.35
15	0.5
20	0.6
25	0.75
30	0.85
45	1.25
60	1.53
80	1.8
100	2.0

533-01-05 Where a circuit-breaker may be operated by a person other than a skilled or instructed person, it shall be designed or installed so that it is not possible to modify the setting or the calibration of its overcurrent release without a deliberate act involving the use of either a key or a tool and resulting in a visible indication of its setting or calibration.

533-01-06 Where necessary to prevent danger, the characteristics and setting of a device for overcurrent protection shall be such that any intended discrimination in its operation is achieved.

533-02 Selection of a device for the protection of a wiring system against overload

533-02-01 The nominal current (or current setting) of the protective device shall be chosen in accordance with Regulation 433-02. In certain cases, to avoid unintentional operation, the peak current values of the loads may have to be taken into consideration.

In the case of a cyclic load, the values of I_n and I_2 shall be chosen on the basis of values of I_b and I_z for the thermally equivalent constant load.

Where the relevant symbols are defined as:

I_b the current for which the circuit is designed i.e. the current intended to be carried in normal service

I_z the current-carrying capacity of the cable for continuous service under the particular installation conditions concerned

I_n the nominal current of the overcurrent protective device

I_2 the current giving effective operation of the overload protective device.

533-03 Selection of a device for protection of a wiring system against fault current

533-03-01 The application of the rules of Chapter 43 shall take into account minimum and maximum fault current conditions.

534 (Reserved for future use)

535 DEVICES FOR PROTECTION AGAINST UNDERVOLTAGE

535-01-01 A device for protection against undervoltage shall be selected and erected so as to allow compliance with requirements of Chapter 45.

536 *(Reserved for future use)*

537 ISOLATING AND SWITCHING DEVICES

537-01 General

537-01-01 Isolating and switching devices shall comply with the appropriate requirements of Regulations 537-02 to 537-05. A common device may be used for more than one of these functions if the appropriate requirements for each function are met.

537-02 Devices for isolation

537-02-01 The isolating distance between contacts or other means of isolation when in the open position shall be not less than that determined for an isolator (disconnector) in accordance with the requirement of BS EN 60 947-3.

537-02-02 A semiconductor device shall not be used as an isolator (disconnector).

537-02-03 The position of the contacts or other means of isolation shall be either externally visible or clearly and reliably indicated. An indication of the isolated position shall occur only when the specified isolating distance has been attained in each pole.

537-02-04 For a 4-wire three-phase supply, where a link is inserted in the neutral conductor, the link shall comply with either or both of the following requirements:

(i) it cannot be removed without the use of tools

(ii) it is accessible to skilled persons only.

537-02-05 A device for isolation shall be selected and/or installed in such a way as to prevent unintentional reclosure, such as that which may be caused by mechanical shock or vibration.

537-02-06 Provision shall be made for securing an off-load isolation device against inadvertent and unauthorised operation.

537-02-07 Multipole isolation shall be achieved by the use of a single device disconnecting the appropriate poles of the supply or alternatively by the use of single-pole devices which are situated adjacent to each other.

537-03 Devices for switching off for mechanical maintenance

537-03-01 A device for switching off for mechanical maintenance shall be inserted where practicable in the main supply circuit. Alternatively, such a device may be inserted in the control circuit, provided that supplementary precautions are taken to provide a degree of safety equivalent to that of interruption of the main supply, e.g. where such an arrangement is specified in the appropriate British Standard.

537-03-02 A device for switching off for mechanical maintenance, or a control switch for such a device, shall be manually initiated and shall have an externally visible contact gap or a clearly and reliably indicated OFF or OPEN position. Indication of that position shall occur only when the OFF or OPEN position on each pole has been fully attained.

537-03-03 A device for switching off for mechanical maintenance shall be selected and/or installed in such a way as to prevent unintentional reclosure, such as that which may be caused by mechanical shock or vibration.

537-03-04 Where a switch is used as a device for switching off for mechanical maintenance, it shall be capable of cutting off the full load current of the relevant part of the installation.

537-04 Devices for emergency switching

537-04-01 A means of interrupting the supply for the purpose of emergency switching shall be capable of cutting off the full load current of the relevant part of the installation. Where appropriate, due account shall be taken of stalled motor conditions.

537-04-02 Means for emergency switching shall consist of:

(i) a single switching device directly cutting off the incoming supply, or

(ii) a combination of several items of equipment operated by a single action and resulting in the removal of the hazard by cutting off the appropriate supply; emergency stopping may include the retention of supply for electric braking facilities.

A plug and socket-outlet shall not be selected as a device for emergency switching.

537-04-03 Where practicable a device for emergency switching shall be manually operated directly interrupting the main circuit. A device such as a circuit-breaker or a contactor operated by remote control shall open on de-energisation of the coil, or another technique of suitable reliability shall be employed.

537-04-04 The operating means (such as handle or pushbutton) for a device for emergency switching shall be clearly identifiable and preferably coloured red. It shall be installed in a readily accessible position where the hazard might occur and, where appropriate, further devices shall be provided where additional emergency switching may be needed.

537-04-05 The operating means of the device for emergency switching shall be of the latching type or capable of being restrained in the OFF or STOP position. A device in which the operating means automatically resets is permitted where both that operating means and the means of re-energising are under the control of one and the same person. The release of the emergency switching device shall not re-energise the equipment concerned.

537-04-06 A fireman's emergency switch provided for compliance with Regulations 476-03-05 to 476-03-07 shall:

(i) be coloured red and have fixed on or near it a permanent durable nameplate marked with the words 'FIREMAN'S SWITCH' the plate being the minimum size 150 mm by 100 mm, in lettering easily legible from a distance appropriate to the site conditions but not less than 36 point, and

(ii) have its ON and OFF positions clearly indicated by lettering legible to a person standing on the ground at the intended site, with the OFF position at the top, and

(iii) be provided with a device to prevent the switch being inadvertently returned to the ON position, and

(iv) be arranged to facilitate operation by a fireman.

537-05   Devices for functional switching

537-05-01 A plug and socket-outlet of rating not exceeding 16 A may be used as a switching device.

537-05-02 Excepting for use on d.c. where this purpose is specifically excluded, a plug and socket-outlet of rating exceeding 16 A may be used as a switching device where the plug and socket-outlet has a breaking capacity appropriate to the use intended (see also Regulation 537-04-02 regarding emergency switching).

CHAPTER 54

EARTHING ARRANGEMENTS AND PROTECTIVE CONDUCTORS

CONTENTS

541	GENERAL
542	CONNECTIONS TO EARTH
542-01	**Earthing arrangements**
542-02	**Earth electrodes**
542-03	**Earthing conductors**
542-04	**Main earthing terminals or bars**
543	PROTECTIVE CONDUCTORS
543-01	**Cross-sectional areas**
543-02	**Types of protective conductor**
543-03	**Preservation of electrical continuity of protective conductors**
544	*(Reserved for future use)*
545	EARTHING ARRANGEMENTS FOR FUNCTIONAL PURPOSES
546	EARTHING ARRANGEMENTS FOR COMBINED PROTECTIVE AND FUNCTIONAL PURPOSES
546-01	**General**
546-02	**Combined protective and neutral (PEN) conductors**
547	PROTECTIVE BONDING CONDUCTORS
547-01	**General**
547-02	**Main equipotential bonding conductors**
547-03	**Supplementary bonding conductors**

CHAPTER 54

EARTHING ARRANGEMENTS AND PROTECTIVE CONDUCTORS

541 GENERAL

541-01-01 Every means of earthing and every protective conductor shall be selected and erected so as to satisfy the requirements of the Regulations.

541-01-02 The earthing system of the installation may be subdivided, in which case each part thus divided shall comply with the requirements of this Chapter.

541-01-03 Where, in an installation, there is also a lightning protection system due account shall be taken of the requirements prescribed in BS 6651. Where bonding to a lightning protection system is provided in accordance with the recommendations of BS 6651, it shall be of no greater cross-sectional area than the earthing conductor.

542 CONNECTIONS TO EARTH

542-01 Earthing arrangements

542-01-01 The main earthing terminal shall be connected with Earth by one of the methods described in Regulations 542-01-02 to 542-01-05, as appropriate to the type of system of which the installation is to form a part (see Section 312 and definition of a 'System') and in compliance with Regulations 542-01-06 to 542-01-08.

542-01-02 For a TN-S system, means shall be provided for the main earthing terminal of the installation to be connected to the earthed point of the source of energy. Part of the connection may be formed by the supplier's lines and equipment.

542-01-03 For a TN-C-S system, where Protective Multiple Earthing is provided, means shall be provided for the main earthing terminal of the installation to be connected by the supplier to the neutral of the source of energy.

542-01-04 For a TT or IT system, the main earthing terminal shall be connected via an earthing conductor to an earth electrode complying with Regulation 542-02.

542-01-05 For a TN-C system, means shall be provided for the connection of PEN conductors to the main earthing terminal.

542-01-06 Where the earthing arrangements are used jointly for protective and functional purposes, according to the requirements of the installation, the requirements for protective measures shall take precedence over any functional requirements.

542-01-07 The earthing arrangements shall be such that:

(i) the value of impedance from the consumer's main earthing terminal to the earthed point of the supply for TN systems, or to Earth for TT and IT systems, is in accordance with the protective and functional requirements of the installation, and considered to be continuously effective, and

(ii) earth fault currents and earth leakage currents which may occur are carried without danger, particularly from thermal, thermomechanical and electromechanical stresses, and

(iii) they are adequately robust or have additional mechanical protection appropriate to the assessed conditions of external influence.

542-01-08 Precautions shall be taken against the risk of damage to other metallic parts through electrolysis.

542-01-09 Where a number of installations have separate earthing arrangements, any protective conductors common to any of these installations shall either be capable of carrying the maximum fault current likely to flow through them, or be earthed within one installation only and insulated from the earthing arrangements of any other installation. In the latter circumstances, if the protective conductor forms part of a cable, the protective conductor shall be earthed only in the installation containing the associated protective device.

542-02 Earth electrodes

542-02-01 The following types of earth electrode are recognised for the purposes of the Regulations:

(i) earth rods or pipes,

(ii) earth tapes or wires,

(iii) earth plates,

(iv) underground structural metalwork embedded in foundations,

(v) welded metal reinforcement of concrete (except pre-stressed concrete) embedded in the earth,

(vi) lead sheaths and other metal coverings of cables, where not precluded by Regulation 542-02-05,

(vii) other suitable underground metalwork.

542-02-02 The type and embedded depth of an earth electrode shall be such that soil drying and freezing will not increase its resistance above the required value.

542-02-03 The design used, and the construction of, an earth electrode shall be such as to withstand damage and to take account of possible increase in resistance due to corrosion.

542-02-04 The metalwork of a gas, water or other service shall not be used as a protective earth electrode. This requirement does not preclude the bonding of such metalwork as required by Regulation 413-02.

542-02-05 The use, as an earth electrode, of the lead sheath or other metal covering of cable shall be subject to all of the following conditions:

(i) the sheath or covering shall be in effective contact with earth,

(ii) the consent of the owner of the cable shall be obtained, and

(iii) arrangements shall exist for the owner of the electrical installation to be warned of any proposed change to the cable which might affect its suitability as an earth electrode.

542-03 Earthing conductors

542-03-01 Every earthing conductor shall comply with Section 543 and in addition, where buried in the ground, shall have a cross-sectional area not less than that stated in Table 54A. For a tape or strip conductor, the thickness shall be such as to withstand mechanical damage and corrosion (see CP 1013).

TABLE 54A

Minimum cross-sectional areas of a buried earthing conductor

	Protected against mechanical damage	Not protected against mechanical damage
Protected against corrosion by a sheath	as required by Regulation 543-01	16 mm^2 copper 16 mm^2 coated steel
Not protected against corrosion	25 mm^2 copper 50 mm^2 steel	25 mm^2 copper 50 mm^2 steel

542-03-02 Neither an aluminium nor a copperclad aluminium conductor shall be used for underground connection to an earth electrode.

542-03-03 The connection of an earthing conductor to an earth electrode or other means of earthing shall be soundly made and be electrically and mechanically satisfactory, and labelled in accordance with Regulation 514-13-01. It shall be suitably protected against corrosion.

542-04 Main earthing terminals or bars

542-04-01 In every installation a main earthing terminal shall be provided to connect the following to the earthing conductor:

  (i) the circuit protective conductors, and

  (ii) the main bonding conductors, and

  (iii) functional earthing conductors (if required), and

  (iv) lightning protection system bonding conductor (if any).

542-04-02 Provision shall be made, in an accessible position, for disconnecting the earthing conductor, to permit measurement of the resistance of the means of earthing when it is part of the installation. This joint shall be such that it can be disconnected only by means of a tool, is mechanically strong and will reliably maintain electrical continuity.

543 PROTECTIVE CONDUCTORS

543-01 Cross-sectional areas

543-01-01 The cross-sectional area of every protective conductor, other than an equipotential bonding conductor, shall be:

  (i) calculated in accordance with Regulation 543-01-03, or

  (ii) selected in accordance with Regulation 543-01-04.

Calculation in accordance with Regulation 543-01-03 is necessary if the choice of cross-sectional areas of phase conductors has been determined by considerations of short-circuit current and if the earth fault current is expected to be less than the short-circuit current.

If the protective conductor:

  (iii) is not an integral part of a cable, or

  (iv) is not formed by conduit, ducting or trunking, or

  (v) is not contained in an enclosure formed by a wiring system,

the cross-sectional area shall be not less than 2.5 mm^2 copper equivalent (see also Regulation 543-03-01).

For an earthing conductor buried in the ground Regulation 542-03-01 also applies.

The cross-sectional area of an equipotential bonding conductor shall comply with Section 547.

543-01-02 Where a protective conductor is common to several circuits, the cross-sectional area of the protective conductor shall be:

  (i) calculated in accordance with Regulation 543-01-03 for the most onerous of the values of fault current and operating time encountered in each of the various circuits, or

  (ii) selected in accordance with Regulation 543-01-04 so as to correspond to the cross-sectional area of the largest phase conductor of the circuits.

543-01-03 The cross-sectional area, where calculated, shall be not less than the value determined by the following formula or shall be obtained by reference to BS 7454.

$$S = \frac{\sqrt{I^2 t}}{k}$$

Where:

  S is the nominal cross-sectional area of the conductor in mm^2.

  I is the value in amperes (r.m.s. for a.c.) of fault current for a fault of negligible impedance, which can flow through the associated protective device, due account being taken of the current limiting effect of the circuit impedances and the limiting capability ($I^2 t$) of that protective device.

Account shall be taken of the effect, on the resistance of circuit conductors, of their temperature rise during the clearance of the fault.

t is the operating time of the disconnecting device in seconds corresponding to the fault current I amperes.

k is a factor taking account of the resistivity, temperature coefficient and heat capacity of the conductor material, and the appropriate initial and final temperatures.

Values of k for protective conductors in various use or service are as given in Tables 54B, 54C, 54D, 54E and 54F. The values are based on the initial and final temperatures indicated below each table.

Where the application of the formula produces a non-standard size, a conductor of the nearest larger standard cross-sectional area shall be used.

TABLE 54B

Values of k for insulated protective conductor not incorporated in a cable and not bunched with cables, or for separate bare protective conductor in contact with cable covering but not bunched with cables, where the assumed initial temperature is $30^\circ C$

Material of conductor	Insulation of protective conductor or cable covering			
	$70^\circ C$ p.v.c.	$85^\circ C$ p.v.c.	$85^\circ C$ rubber	$90^\circ C$ thermosetting
Copper	143/133*	143/133*	166	176
Aluminium	95/88*	95/88*	110	116
Steel	52	52	60	64
Assumed initial temperature	$30^\circ C$	$30^\circ C$	$30^\circ C$	$30^\circ C$
Final temperature	$160^\circ C/140^\circ C$*	$160^\circ C/140^\circ C$*	$220^\circ C$	$250^\circ C$

* Above 300 mm^2

TABLE 54C

Values of k for protective conductor incorporated in a cable or bunched with cables, where the assumed initial temperature is $70^\circ C$ or greater.

Material of conductor	Insulation Material			
	$70^\circ C$ p.v.c.	$85^\circ C$ p.v.c.	$85^\circ C$ rubber	$90^\circ C$ thermosetting
Copper	115/103*	104/90*	134	143
Aluminium	76/68*	69/60*	89	94
Assumed initial temperature	$70^\circ C$	$85^\circ C$	$85^\circ C$	$90^\circ C$
Final temperature	$160^\circ C/140^\circ C$*	$160^\circ C/140^\circ C$*	$220^\circ C$	$250^\circ C$

* Above 300 mm^2

TABLE 54D

Values of k for protective conductor as a sheath or armour of a cable

Material of conductor	Insulation Material			
	70°C p.v.c.	85°C p.v.c.	85°C rubber	90°C thermosetting
Aluminium	93	87	93	85
Steel	51	48	51	46
Lead	26	24	26	23
Assumed initial temperature	60°C	75°C	75°C	80°C
Final temperature	200°C	200°C	220°C	200°C

TABLE 54E

Values of k for steel conduit, ducting and trunking as the protective conductor

Material of protective conductor conduit	Insulation Material			
	70°C p.v.c.	85°C p.v.c.	85°C rubber	90°C thermosetting
Steel conduit, ducting and trunking	47	45	54	58
Assumed initial temperature	50°C	58°C	58°C	60°C
Final temperature	160°C	160°C	220°C	250°C

TABLE 54F

Values of k for bare conductor where there is no risk of damage to any neighbouring material by the temperatures indicated

The temperatures indicated are valid only where they do not impair the quality of the connections

Material of conductor	Conditions		
	Visible and in restricted areas	Normal conditions	Fire risk
Copper	228	159	138
Aluminium	125	105	91
Steel	82	58	50
Assumed initial temperature	30°C	30°C	30°C
Final temperature			
Copper conductors	500°C	200°C	150°C
Aluminium conductors	300°C	200°C	150°C
Steel conductors	500°C	200°C	150°C

543-01-04 Where it is desired not to calculate the minimum cross-sectional area of a protective conductor in accordance with Regulation 543-01-03, the cross-sectional area may be determined in accordance with Table 54G.

Where the application of Table 54G produces a non-standard size, a conductor having the nearest larger standard cross-sectional area shall be used.

TABLE 54G

Minimum cross-sectional area of protective conductor in relation to the cross-sectional area of associated phase conductor

Cross-sectional area of phase conductor (S)	Minimum cross-sectional area of the corresponding protective conductor (Sp)	
	If the protective conductor is of the same material as the phase conductor	If the protective conductor is not the same material as the phase conductor
mm^2	mm^2	mm^2
$S \leqslant 16$	S	$\dfrac{k_1 S}{k_2}$
$16 < S \leqslant 35$	16	$\dfrac{k_1 16}{k_2}$
$S > 35$	$\dfrac{S}{2}$	$\dfrac{k_1 S}{k_2 2}$

Where:

k_1 is the value of k for the phase conductor, selected from TABLE 43A in Chapter 43 according to the materials of both conductor and insulation.

k_2 is the value of k for the protective conductor, selected from TABLES 54B, 54C, 54D, 54E, or 54F as applicable.

543-02   Types of protective conductor

543-02-01   Flexible or pliable conduit shall not be selected as a protective conductor. Neither a gas pipe nor an oil pipe shall be selected as a protective conductor.

543-02-02   A protective conductor may consist of one or more of the following:

(i)   a single-core cable

(ii)   a conductor in a cable

(iii)   an insulated or bare conductor in a common enclosure with insulated live conductors

(iv)   a fixed bare or insulated conductor

(v)   a metal covering, for example, the sheath, screen or armouring of a cable

(vi)   a metal conduit or other enclosure or electrically continuous support system for conductors

(vii)   an extraneous-conductive-part complying with Regulation 543-02-06.

543-02-03   A protective conductor of the types described in Items (i) to (iv) of Regulation 543-02-02 and of cross-sectional area 10 mm^2 or less, shall be of copper.

543-02-04   Where a metal enclosure or frame of a low voltage switchgear or controlgear assembly or busbar trunking system is used as a protective conductor, it shall satisfy the following three requirements:

(i)   its electrical continuity shall be assured, either by construction or by suitable connection, in such a way as to be protected against mechanical, chemical or electrochemical deterioration, and

(ii)   its cross-sectional area shall be at least equal to that resulting from the application of Regulation 543-01, or verified by test in accordance with BS 5486 Part 1, and

(iii)   it shall permit the connection of other protective conductors at every predetermined tap-off point.

543-02-05 The metal covering including the sheath (bare or insulated) of a cable, in particular the sheath of a mineral-insulated cable, trunking and ducting for electrical purposes and metal conduit, may be used as a protective conductor for the associated circuit, if it satisfies both requirements of Items (i) and (ii) of Regulation 543-02-04.

543-02-06 Except as prohibited in Regulation 543-02-01 an extraneous-conductive-part may be used as a bonding conductor. Suitable structural metalwork may be used as a protective conductor. In either case both of the following requirements shall be satisfied:

(i) the electrical continuity shall be assured, either by construction or by suitable connection, in such a way as to be protected against mechanical, chemical or electrochemical deterioration

(ii) the cross-sectional area shall be at least equal to that resulting from the application of Regulation 543-01.

543-02-07 Where the protective conductor is formed by conduit, trunking, ducting, or the metal sheath and/or armour of a cable, the earthing terminal of each accessory shall be connected by a separate protective conductor to an earthing terminal incorporated in the associated box or other enclosure.

543-02-08 An exposed-conductive-part of equipment shall not be used to form a protective conductor for other equipment except as provided by Regulations 543-02-02, 543-02-04 and 543-02-05.

543-02-09 Except where the circuit protective conductor is formed by a metal covering or enclosure containing all of the conductors of the ring, the circuit protective conductor of every ring final circuit shall also be run in the form of a ring having both ends connected to the earthing terminal at the origin of the circuit.

543-02-10 A separate metal enclosure for cable shall not be used as a PEN conductor.

543-03 Preservation of electrical continuity of protective conductors

543-03-01 A protective conductor shall be suitably protected against mechanical and chemical deterioration and electrodynamic effects.

543-03-02 Excepting (i) and (ii) below a protective conductor having a cross-sectional area up to and including 6 mm^2 shall be protected throughout by a covering at least equivalent to that provided by the insulation of a single-core non-sheathed cable of appropriate size complying with BS 6004 or BS 7211:

(i) a protective conductor forming part of a multicore cable.

(ii) cable trunking or conduit used as a protective conductor.

Where the sheath of a cable incorporating an uninsulated protective conductor of cross-sectional area up to and including 6 mm^2 is removed adjacent to joints and terminations, the protective conductor shall be protected by insulating sleeving complying with BS 2848.

543-03-03 Except for a joint in metal conduit, ducting, trunking or support systems, the connection of a protective conductor shall comply with the requirements for accessibility of Regulation 526-04.

543-03-04 No switching device shall be inserted in a protective conductor except for the following:

(i) as permitted by Regulation 460-01-04

(ii) multipole linked switching or plug-in devices in which the protective conductor circuit shall not be interrupted before the live conductors and shall be re-established not later than when the live conductors are re-connected.

Joints which can be disconnected for test purposes are permitted in a protective conductor circuit.

543-03-05 Where electrical earth monitoring is used, the operating coil shall be connected in the pilot conductor and not in the protective earthing conductor. See BS 4444.

543-03-06 Every joint in metallic conduit shall be mechanically and electrically continuous by screwing or by substantial mechanical clamps. Plain slip or pin-grip sockets shall not be used.

544 *(Reserved for future use)*

545 EARTHING ARRANGEMENTS FOR FUNCTIONAL PURPOSES

545-01-01 Where a 'clean' (low-noise) earth is specified for a particular item of equipment, the manufacturer of the equipment shall be consulted in order to confirm that the arrangements described in Section 607 are suitable for functional purposes.

546 EARTHING ARRANGEMENTS FOR COMBINED PROTECTIVE AND FUNCTIONAL PURPOSES

546-01 General

546-01-01 Where earthing for combined protective and functional purposes is required, the requirements for protective measures shall take precedence.

546-02 Combined protective and neutral (PEN) conductors

546-02-01 The provisions of Regulations 546-02-02 to 546-02-08 may be applied only:

(i) where any necessary authorisation for use of a PEN conductor has been obtained by the supplier and where the installation complies with the conditions for that authorisation, or

(ii) where the installation is supplied by a privately owned transformer or convertor in such a way that there is no metallic connection (except for the earthing connection) with the general public supply, or

(iii) where the supply is obtained from a private generating plant.

546-02-02 A conductor of the following types may serve as a PEN conductor provided that the part of the installation concerned is not supplied through a residual current device:

(i) for a fixed installation, a conductor of a cable not subject to flexing and having a cross-sectional area not less than 10 mm^2 for copper, or 16 mm^2 for aluminium

(ii) the outer conductor of a concentric cable where that conductor has a cross-sectional area not less than 4 mm^2 in a cable complying with an appropriate British Standard and selected and erected in accordance with Regulations 546-02-03 to 546-02-08.

546-02-03 The outer conductor of a concentric cable shall not be common to more than one circuit. This requirement does not preclude the use of a twin or multicore cable to serve a number of points contained within one final circuit.

546-02-04 The conductance of the outer conductor of a concentric cable (measured at a temperature of 20°C) shall:

(i) for a single-core cable, be not less than that of the internal conductor

(ii) excepting for a cable complying with BS 5593, for a multicore cable in a multiphase or multipole circuit, be not less than that of one internal conductor

(iii) for a multicore cable serving a number of points contained within one final circuit or having the internal conductors connected in parallel, be not less than that of the internal conductors connected in parallel.

546-02-05 At every joint in the outer conductor of a concentric cable and at a termination, the continuity of that joint shall be supplemented by a conductor additional to any means used for sealing and clamping the outer conductor. The conductance of the additional conductor shall be not less than that specified in Regulation 546-02-04 for the outer conductor.

546-02-06 No means of isolation or switching shall be inserted in the outer conductor of a concentric cable.

546-02-07 Excepting a cable to BS 6207 installed in accordance with manufacturers' instructions, the PEN conductor of every cable shall be insulated or have an insulating covering suitable for the highest voltage to which it may be subjected.

546-02-08 If from any point of the installation the neutral and protective functions are provided by separate conductors, those conductors shall not then be re-connected together beyond that point. At the point of separation, separate terminals or bars shall be provided for the protective and neutral conductors. The PEN conductor shall be connected to the terminals or bar intended for the protective earthing conductor and the neutral conductor. The conductance of the terminal link or bar shall not be less than that specified in Regulation 546-02-04.

547 PROTECTIVE BONDING CONDUCTORS

547-01 General

547-01-01 An aluminium or copperclad aluminium conductor shall not be used for a bonding connection.

547-02 Main equipotential bonding conductors

547-02-01 Except where PME conditions apply a main equipotential bonding conductor shall have a cross-sectional area not less than half the cross-sectional area required for the earthing conductor of the installation and not less than 6 mm^2. The cross-sectional area need not exceed 25 mm^2 if the bonding conductor is of copper or a cross-sectional area affording equivalent conductance in other metals.

Where PME conditions apply the main equipotential bonding conductor shall be selected in accordance with the neutral conductor of the supply and Table 54H.

TABLE 54H

Minimum cross-sectional area of the main equipotential bonding conductor in relation to the neutral of the supply

Copper equivalent cross-sectional area of the supply neutral conductor	Minimum copper equivalent cross-sectional area of the main equipotential bonding conductor
35 mm^2 or less	10 mm^2
over 35 mm^2 up to 50 mm^2	16 mm^2
over 50 mm^2 up to 95 mm^2	25 mm^2
over 95 mm^2 up to 150 mm^2	35 mm^2
over 150 mm^2	50 mm^2

547-02-02 The main equipotential bonding connection to any gas, water or other service shall be made as near as practicable to the point of entry of that service into the premises.

Where there is an insulating section or insert at that point, or there is a meter, the connection shall be made to the consumer's hard metal pipework and before any branch pipework. Where practicable the connection shall be made within 600 mm of the meter outlet union or at the point of entry to the building if the meter is external.

547-03 Supplementary bonding conductors

547-03-01 A supplementary bonding conductor connecting two exposed-conductive-parts shall have a conductance, if sheathed or otherwise provided with mechanical protection, not less than that of the smaller protective conductor connected to the exposed-conductive-parts. If mechanical protection is not provided, its cross-sectional area shall be not less than 4 mm^2.

547-03-02 A supplementary bonding conductor connecting an exposed-conductive-part to an extraneous-conductive-part shall have a conductance, if sheathed or otherwise provided with mechanical protection, not less than half that of the protective conductor connected to the exposed-conductive-part. If mechanical protection is not provided, its cross-sectional area shall be not less than 4 mm^2.

547-03-03 A supplementary bonding conductor connecting two extraneous-conductive-parts shall have a cross-sectional area not less than 2.5 mm^2 if sheathed or otherwise provided with mechanical protection or 4 mm^2 if mechanical protection is not provided, except that where one of the extraneous-conductive-parts is connected to an exposed-conductive-part in compliance with Regulation 547-03-02, that Regulation shall apply also to the conductor connecting the two extraneous-conductive-parts.

547-03-04 Except where Regulation 547-03-05 applies, supplementary bonding shall be provided by a supplementary conductor, a conductive part of a permanent and reliable nature, or by a combination of these.

547-03-05 Where supplementary bonding is to be applied to a fixed appliance which is supplied via a short length of flexible cord from an adjacent connection unit or other accessory, incorporating a flex outlet, the circuit protective conductor within the flexible cord shall be deemed to provide the supplementary bonding connection to the exposed-conductive-parts of the appliance, from the earthing terminal in the connection unit or other accessory.

CHAPTER 55

OTHER EQUIPMENT

CONTENTS

551	TRANSFORMERS
551-01	**Autotransformers and step-up transformers**
552	ROTATING MACHINES
552-01	**Rotating machines**
553	ACCESSORIES
553-01	**Plugs and socket-outlets**
553-02	**Cable couplers**
553-03	**Lampholders**
553-04	**Lighting points**
554	CURRENT-USING EQUIPMENT
554-01	**Luminaires**
554-02	**High voltage discharge lighting installations**
554-03	**Electrode water heaters and boilers**
554-04	**Heaters for liquids or other substances having immersed heating elements**
554-05	**Water heaters having immersed and uninsulated heating elements**
554-06	**Heating conductors and cables**
554-07	**Electric surface heating systems**

CHAPTER 55

OTHER EQUIPMENT

551 TRANSFORMERS

551-01 Autotransformers and step-up transformers

551-01-01 Where an autotransformer is connected to a circuit having a neutral conductor, the common terminal of the winding shall be connected to the neutral conductor.

551-01-02 A step-up autotransformer shall not be connected to an IT system.

551-01-03 Where a step-up transformer is used, a linked switch shall be provided for disconnecting the transformer from all live conductors of the supply.

552 ROTATING MACHINES

552-01-01 All equipment, including cable, of every circuit carrying the starting, accelerating and load currents of a motor shall be suitable for a current at least equal to the full-load current rating of the motor when rated in accordance with the appropriate British Standard. Where the motor is intended for intermittent duty and for frequent starting and stopping, account shall be taken of any cumulative effects of the starting or braking currents upon the temperature rise of the equipment of the circuit.

552-01-02 Every electric motor having a rating exceeding 0.37 kW shall be provided with control equipment incorporating means of protection against overload of the motor. This requirement does not apply to a motor incorporated in an item of current-using equipment complying as a whole with an appropriate British Standard.

552-01-03 Except where failure to start after a brief interruption would be likely to cause greater danger, every motor shall be provided with means to prevent automatic restarting after a stoppage due to drop in voltage or failure of supply, where unexpected restarting of the motor might cause danger. These requirements do not preclude arrangements for starting a motor at intervals by an automatic control device, where other adequate precautions are taken against danger from unexpected restarting.

553 ACCESSORIES

553-01 Plugs and socket-outlets

553-01-01 Every plug and socket-outlet shall comply with all the requirements of Items (i) and (ii) below, and in addition shall comply with the appropriate requirements of Regulations 553-01-02 to 553-02-02:

(i) except for SELV circuits, it shall not be possible for any pin of a plug to make contact with any live contact of its associated socket-outlet while any other pin of the plug is completely exposed

(ii) it shall not be possible for any pin of a plug to make contact with any live contact of any socket-outlet within the same installation other than the type of socket-outlet for which the plug is designed.

553-01-02 Except for SELV or a special circuit from Regulation 553-01-05, every plug and socket-outlet shall be of the non-reversible type, with provision for the connection of a protective conductor.

553-01-03 Except where Regulation 553-01-05 applies, in a low voltage circuit every plug and socket-outlet shall conform with the applicable British Standard listed in Table 55A.

TABLE 55A

Plugs and socket-outlets for low voltage circuits

Type of plug and socket-outlet	Rating (amperes)	Applicable British Standard
Fused plugs and shuttered socket-outlets, 2-pole and earth, for a.c.	13	BS 1363 (fuses to BS 1362)
Plugs, fused or non-fused, and socket-outlets, 2-pole and earth	2,5,15,30	BS 546 (fuses, if any, to BS 646)
Plugs, fused or non-fused, and socket-outlets, protected type, 2-pole with earthing contact	5,15,30	BS 196
Plugs and socket-outlets (industrial type)	16,32,63,125	BS 4343

553-01-04 Every socket-outlet for household and similar use shall be of the shuttered type and, for an a.c. installation, shall preferably be of a type complying with BS 1363.

553-01-05 A plug and socket-outlet not complying with BS 1363, BS 546, BS 196 or BS 4343, may be used in single-phase a.c. or two wire d.c. circuits operating at a nominal voltage not exceeding 250 volts for:

(i) the connection of an electric clock, provided that the plug and socket-outlet are designed specifically for that purpose, and that each plug incorporates a fuse of rating not exceeding 3 amperes complying with BS 646 or BS 1362 as appropriate

(ii) the connection of an electric shaver, provided that the socket-outlet is either incorporated in a shaver supply unit complying with BS 3535 or, in a room other than a bathroom, is a type complying with BS 4573

(iii) a circuit having special characteristics such that danger would otherwise arise or it is necessary to distinguish the function of the circuit.

553-01-06 A socket-outlet on a wall or similar structure shall be mounted at a height above the floor or any working surface to minimise the risk of mechanical damage to the socket-outlet or to an associated plug and its flexible cord which might be caused during insertion, use or withdrawal of the plug.

553-01-07 Where portable equipment is likely to be used, provision shall be made so that the equipment can be fed from an adjacent and conveniently accessible socket-outlet, taking account of the length of flexible cord normally fitted to portable appliances and luminaires.

553-02 Cable couplers

553-02-01 Except for a SELV or a Class II circuit, a cable coupler shall comply where appropriate with BS 196, BS 4343, BS 4491, or BS 6991, shall be non-reversible and shall have provision for the connection of a protective conductor.

553-02-02 A cable coupler shall be arranged so that the connector of the coupler is fitted at the end of the cable remote from the supply.

553-03 Lampholders

553-03-01 Except where the lampholder and its wiring are enclosed in earthed metal or insulating material having the ignitability characteristic 'P' as specified in BS 476 Part 5 or where separate overcurrent protection is provided, a lampholder shall not be connected to any circuit where the rated current of the overcurrent protective device exceeds the appropriate value stated in Table 55B.

TABLE 55B

Overcurrent protection of lampholders

Type of lampholder (as designated in BS 5042, BS 6776 and BS 6702)			Maximum rating (amperes) of overcurrent protective device protecting the circuit
Bayonet (BS 5042) :	B15	SBC	6
	B22	BC	16
Edison screw (BS 6776):	E14	SES	6
	E27	ES	16
	E40	GES	16

553-03-02 A lampholder for a filament lamp shall not be installed in a circuit operating at a voltage exceeding 250 volts.

553-03-03 Bayonet lampholders B15 and B22 shall comply with BS 5042 and shall have the temperature rating T2 described in that British Standard.

553-03-04 For a circuit of a TN or a TT system, the outer contact of every Edison screw or single centre bayonet cap type lampholder shall be connected to the neutral conductor. This regulation applies equally to track mounted systems.

553-04 Lighting points

553-04-01 At each fixed lighting point one of the following accessories shall be used:

(i) a ceiling rose to BS 67

(ii) a luminaire supporting coupler to BS 6972 or BS 7001

(iii) a batten lampholder to BS 5042 or BS 6776

(iv) a luminaire designed to be connected directly to the circuit wiring.

A lighting accessory or luminaire shall be controlled by a switch or combination of switches to BS 3676 and/or BS 5518 and where appropriate shall be suitable for discharge lighting circuits.

553-04-02 A ceiling rose shall not be installed in any circuit operating at a voltage normally exceeding 250 volts.

553-04-03 A ceiling rose shall not be used for the attachment of more than one outgoing flexible cord unless it is specially designed for multiple pendants.

553-04-04 Luminaire supporting couplers are designed specifically for the mechanical support and electrical connection of luminaires and shall not be used for the connection of any other equipment.

554 CURRENT-USING EQUIPMENT

554-01 Luminaires

554-01-01 Where a pendant luminaire is installed, the associated accessory shall be suitable for the mass suspended (see also Regulation 522-08-06).

554-01-02 An extra-low voltage luminaire without provision for the connection of a protective conductor shall be installed only as part of a SELV system.

554-02 High voltage discharge lighting installations

554-02-01 Every high voltage electric sign and high voltage luminous discharge tube installation shall be constructed, selected and erected in accordance with the requirements of BS 559.

554-03 Electrode water heaters and boilers

554-03-01 Every electrode boiler and electrode water heater shall be connected to an a.c. system only, and shall be selected and erected in accordance with the appropriate requirements of this Section.

554-03-02 The supply to the heater or boiler shall be controlled by a linked circuit-breaker arranged to disconnect the supply from all electrodes simultaneously and provided with an overcurrent protective device in each conductor feeding an electrode.

554-03-03 The earthing of the heater or boiler shall comply with the requirements of Chapter 54 and, in addition, the shell of the heater or boiler shall be bonded to the metallic sheath and armour, if any, of the incoming supply cable. The protective conductor shall be connected to the shell of the heater or boiler and shall comply with Regulation 543-01-01.

554-03-04 Where an electrode water heater or electrode boiler is directly connected to a supply at a voltage exceeding low voltage, the installation shall include a residual current device arranged to disconnect the supply from the electrodes on the occurrence of a sustained earth-leakage current in excess of 10% of the rated current of the heater or boiler under normal conditions of operation, except that if in any instance a higher value is essential to ensure stability of operation of the heater or boiler, the value may be increased to a maximum of 15%. A time delay may be incorporated in the device to prevent unnecessary operation in the event of imbalance of short duration.

554-03-05 Where an electrode water heater or electrode boiler is connected to a three-phase low-voltage supply, the shell of the heater or boiler shall be connected to the neutral of the supply as well as to the earthing conductor. The current-carrying capacity of the neutral conductor shall be not less than that of the largest phase conductor connected to the equipment.

554-03-06 Except as provided by Regulation 554-03-07, where the supply to an electrode water heater or electrode boiler is single-phase and one electrode is connected to a neutral conductor earthed by the supplier, the shell of the water-heater or boiler shall be connected to the neutral of the supply as well as to the earthing conductor.

554-03-07 Where the heater or boiler is not piped to a water supply or in physical contact with any earthed metal, and where the electrodes and the water in contact with the electrodes are so shielded in insulating material that they cannot be touched while the electrodes are live, a fuse in the phase conductor may be substituted for the circuit-breaker required under Regulation 554-03-02 and the shell of the heater or boiler need not be connected to the neutral of the supply.

554-04 Heaters for liquids or other substances, having immersed heating elements

554-04-01 Every heater for liquid or other substance shall incorporate or be provided with an automatic device to prevent a dangerous rise in temperature.

554-05 Water heaters having immersed and uninsulated heating elements

554-05-01 Every single-phase water heater or boiler having an uninsulated heating element immersed in the water shall comply with the requirements of Regulations 554-05-02 and 554-05-03. This type of water heater or boiler is deemed not to be an electrode water heater or boiler.

554-05-02 All metal parts of the heater or boiler which are in contact with the water (other than current-carrying parts) shall be solidly and metallically connected to a metal water-pipe through which the water supply to the heater or boiler is provided, and that water-pipe shall be connected to the main earthing terminal by means independent of the circuit protective conductor.

554-05-03 The heater or boiler shall be permanently connected to the electricity supply through a double-pole linked switch which is either separate from and within easy reach of the heater or boiler or is incorporated therein and the wiring from the heater or boiler shall be directly connected to that switch without use of a plug and socket-outlet; and, where the heater or boiler is installed in a room containing a fixed bath, the switch shall comply in addition with Section 601.

554-05-04 Before a heater or boiler of the type referred to in Regulation 554-05 is connected, the installer shall confirm that no single-pole switch, non-linked circuit-breaker or fuse is fitted in the neutral conductor in any part of the circuit between the heater or boiler and the origin of the installation.

554-06 Heating conductors and cables

554-06-01 Where a heating cable is required to pass through, or be in close proximity to, material which presents a fire hazard, the cable shall be enclosed in material having the ignitability characteristic 'P' as specified in BS 476, Part 5 and shall be adequately protected from any mechanical damage reasonably foreseeable during installation and use.

554-06-02 A heating cable intended for laying directly in soil, concrete, cement screed or other material used for road and building construction shall be:

(i) capable of withstanding mechanical damage under the conditions that can reasonably be expected to prevail during its installation, and

(ii) constructed of material that will be resistant to damage from dampness and/or corrosion under normal conditions of service.

554-06-03 A heating cable laid directly in soil, a road, or the structure of a building shall be installed so that it:

(i) is completely embedded in the substance it is intended to heat, and

(ii) does not suffer damage in the event of movement normally to be expected in it or the substance in which it is embedded, and

(iii) complies in all respect with maker's instructions and recommendations.

554-06-04 The loading of every floor-warming cable under operating conditions shall be limited to a value such that the appropriate conductor temperature specified in Table 55C is not exceeded.

TABLE 55C

Maximum conductor operating temperatures for a floor-warming cable

Type of cable	Maximum conductor operating temperature $^{\circ}$C
General-purpose p.v.c. over conductor	70
Enamelled conductor, polychlorophene over enamel, p.v.c. overall	70
Enamelled conductor p.v.c. overall	70
Enamelled conductor, p.v.c. over enamel, lead-alloy 'E' sheath overall	70
Heat-resisting p.v.c. over conductor	85
Nylon over conductor, heat-resisting p.v.c. overall	85
Synthetic rubber or equivalent elastomeric insulation over conductor	85
Mineral insulation over conductor, copper sheath overall	*
Silicone-treated woven-glass sleeve over conductor	180

* The temperature depends upon many factors including the following:

(i) the type of seal employed,
(ii) whether the heating section is connected to a cold lead-in section or not,
(iii) the outer covering material, if any, and
(iv) the material in contact with the heating section.

554-07 Electric surface heating systems

554-07-01 The equipment, system design, installation and testing of an electric surface heating (ESH) system shall be in accordance with BS 6351.

CHAPTER 56

SUPPLIES FOR SAFETY SERVICES

CONTENTS

561	GENERAL
562	SOURCES
563	CIRCUITS
564	UTILISATION EQUIPMENT
565	SPECIAL REQUIREMENTS FOR SAFETY SERVICES HAVING SOURCES NOT CAPABLE OF OPERATION IN PARALLEL
566	SPECIAL REQUIREMENTS FOR SAFETY SERVICES HAVING SOURCES CAPABLE OF OPERATION IN PARALLEL

CHAPTER 56

SUPPLIES FOR SAFETY SERVICES

561   **GENERAL**

561-01-01 For a safety service, a source of supply shall be selected which will maintain a supply of adequate duration.

561-01-02 For a safety service required to operate in fire conditions, all equipment shall be provided, either by construction or by erection, with protection providing fire resistance of adequate duration.

561-01-03 A protective measure against indirect contact without automatic disconnection at the first fault is preferred. In an IT system, continuous insulation monitoring shall be provided to give audible and visible indications of a first fault.

562   **SOURCES**

562-01-01 A source for safety services shall be one of the following:

    (i)   a primary cell or cells

    (ii)   a storage battery

    (iii)   a generator set capable of independent operation

    (iv)   a separate feeder effectively independent of the normal feeder (provided that an assessment is made that the two supplies are unlikely to fail concurrently).

562-01-02 A source for a safety service shall be installed as fixed equipment and in such a manner that it cannot be adversely affected by failure of the normal source.

562-01-03 A source for a safety service shall be placed in a suitable location and be accessible only to skilled or instructed persons.

562-01-04 A single source for a safety service shall not be used for another purpose. However, where more than one source is available, such sources may supply standby systems provided that, in the event of failure of one source, the energy remaining available will be sufficient for the starting and operation of all safety services; this generally necessitates the automatic off-loading of equipment not providing safety services.

562-01-05 Regulations 562-01-03 and 562-01-04 do not apply to equipment individually supplied by a self-contained battery.

563   **CIRCUITS**

563-01-01 The circuit of a safety service shall be independent of any other circuit and an electrical fault or any intervention or modification in one system shall not affect the correct functioning of the other.

563-01-02 The circuit of a safety service shall not pass through any location exposed to abnormal fire risk unless the wiring system used is adequately fire resistant.

563-01-03 The protection against overload prescribed in Regulation 473-01-01 may be omitted.

563-01-04 Every overcurrent protective device shall be selected and erected so as to avoid an overcurrent in one circuit impairing the correct operation of any other safety services circuit.

563-01-05 Switchgear and controlgear shall be clearly identified and grouped in locations accessible only to skilled or instructed persons.

563-01-06 Every alarm, indication and control device shall be clearly identified.

564   **UTILISATION EQUIPMENT**

564-01-01 In equipment supplied by two different circuits, a fault occurring in one circuit shall not impair the protection against electric shock nor the correct operation of the other circuit.

565 SPECIAL REQUIREMENTS FOR SAFETY SERVICES HAVING SOURCES NOT CAPABLE OF OPERATION IN PARALLEL

565-01-01 Precautions shall be taken to prevent the paralleling of the sources, e.g. by both mechanical and electrical interlocking.

565-01-02 The requirements of the Regulations for protection against fault current and against indirect contact shall be met for each source.

566 SPECIAL REQUIREMENTS FOR SAFETY SERVICES HAVING SOURCES CAPABLE OF OPERATION IN PARALLEL

566-01-01 The requirements of the Regulations for protection against short-circuit and against indirect contact shall be met whether the installation is supplied by either of the two sources or by both in parallel.

PART 6
SPECIAL INSTALLATIONS OR LOCATIONS - PARTICULAR REQUIREMENTS

CONTENTS

600	GENERAL
601	LOCATIONS CONTAINING A BATH TUB OR SHOWER BASIN
602	SWIMMING POOLS
603	HOT AIR SAUNAS
604	CONSTRUCTION SITE INSTALLATIONS
605	AGRICULTURAL AND HORTICULTURAL PREMISES
606	RESTRICTIVE CONDUCTIVE LOCATIONS
607	EARTHING REQUIREMENTS FOR INSTALLATION OF EQUIPMENT HAVING HIGH EARTH LEAKAGE CURRENTS
608	DIVISION ONE ELECTRICAL INSTALLATIONS IN CARAVANS AND MOTOR CARAVANS
608	DIVISION TWO ELECTRICAL INSTALLATIONS IN CARAVAN PARKS
609	*(Reserved for Marinas)*
610	*(Reserved for future use)*
611	HIGHWAY POWER SUPPLIES AND STREET FURNITURE

600 GENERAL

600-01 The particular requirements for each special installation or location in Part 6 supplement or modify the general requirements contained in other Parts of the Regulations.

600-02 The absence of reference to the exclusion of a Chapter, a Section or a Clause means that the corresponding general Regulations are applicable.

SECTION 601

LOCATIONS CONTAINING A BATH TUB OR SHOWER BASIN

601-01 Scope

601-01-01 The particular requirements of this Section shall apply to bath tubs, shower basins and their surroundings, where the risk of electric shock is increased by a reduction in body resistance and contact of the body with earth potential.

Special requirements may be necessary for a location containing a bath for medical treatment.

Protection for safety

601-02 Protection against electric shock

601-02-01 No electrical equipment shall be installed in the interior of a bath tub or shower basin.

601-03 Protection against both direct and indirect contact

601-03-01 Where SELV (Regulations 411-02 and 471-02) is used, the safety source shall be installed out of reach of a person using the bath or shower and, notwithstanding the provision of the second paragraph of Regulation 411-02-09, the equipment shall incorporate protection against direct contact by insulation (Regulation 412-02), capable of withstanding a test voltage of 500 V r.m.s. a.c. for 60 s, or by barriers or enclosures (Regulation 412-03).

601-04 Protection against indirect contact

601-04-01 Except for SELV, for a circuit supplying equipment in a room containing a fixed bath or shower, where the equipment is simultaneously accessible with exposed-conductive-parts of other equipment or with extraneous-conductive-parts, the characteristics of the protective devices and the earthing arrangements shall be such that, in the event of a fault to earth, disconnection occurs within 0.4 s.

601-04-02 Except for equipment supplied from a SELV circuit, in a room containing a fixed bath or shower, supplementary equipotential bonding shall be provided between simultaneously accessible exposed conductive parts of equipment, between exposed-conductive-parts and simultaneously accessible extraneous-conductive-parts, and between simultaneously accessible extraneous-conductive-parts.

601-04-03 Where electrical equipment is installed in the space below a bath, that space shall be accessible only by the use of a tool and, nevertheless, the requirement of Regulation 601-04-02 shall extend to the interior of that space.

Application of protective measures against electric shock

601-05 Protection against direct contact

601-05-01 The following protective measures against direct contact shall not be used:

(i) protection by means of obstacles (Regulation 412-04)

(ii) protection by placing out of reach (Regulation 412-05).

601-06 Protection against indirect contact

601-06-01 The following protective measures against indirect contact shall not be used:

(i) protection by non-conducting location (Regulation 413-04)

(ii) protection by means of earth-free local equipotential bonding (Regulation 413-05).

Selection and erection of equipment

601-07 Wiring Systems

601-07-01 Surface wiring systems shall not employ metallic conduit or metallic trunking or an exposed metallic cable sheath or an exposed earthing or bonding conductor.

Switchgear and controlgear

601-08 Devices for isolation and switching

601-08-01 Every switch or other means of electrical control or adjustment shall be so situated as to be normally inaccessible to a person using a fixed bath or shower. This requirement does not apply to:

(i) the insulating cords of cord-operated switches which comply with BS 3676

(ii) mechanical actuators, with linkages incorporating insulating components, of remotely operated switches

(iii) controls which comply with the relevant requirements of BS 3456 Section 3.9 (1979) - Stationary instantaneous water heaters

(iv) switches supplied by SELV at a nominal voltage not exceeding 12 V r.m.s. a.c. or d.c.

(v) a shaver supply unit complying with Regulation 601-09-01.

601-08-02 A switch which forms part of a SELV circuit shall have no accessible metal parts.

Other equipment

601-09 Transformers

601-09-01 In a room containing a fixed bath or shower, provision for the connection of an electric shaver shall be only by means of a shaver supply unit complying with BS 3535 or such a unit incorporated in a luminaire. The earthing terminal of the shaver supply unit shall be connected to the protective conductor of the final circuit from which the supply is derived.

Accessories

601-10 Plugs and socket-outlets

601-10-01 Where a socket-outlet which forms part of a SELV circuit is installed in a room containing a fixed bath or shower, the nominal voltage shall not exceed 12 V r.m.s. a.c. or d.c. and the socket-outlet shall have no accessible metallic parts.

601-10-02 Except as permitted by Regulations 601-09-01 and 601-10-01, in a room containing a fixed bath and in a shower room there shall be no socket-outlet and no provision for connecting portable equipment.

601-10-03 Where a shower cubicle is located in a room other than a bathroom or a shower room, any socket-outlet not provided in conformance with Regulations 601-09-01 or 601-10-01 shall be installed at least 2.5 m from the shower cubicle.

Current-using equipment

601-11 Luminaires

601-11-01 In a room containing a fixed bath or shower cubicle, parts of a lampholder within a distance of 2.5 m from the bath or shower cubicle shall be constructed of, or shrouded in, insulating material. Bayonet lampholders Type B22 shall be fitted with a protective shield complying with BS 5042. Alternatively, totally enclosed luminaires shall be used.

601-12 Other fixed equipment

601-12-01 No stationary appliance having heating elements which can be touched shall be installed within reach of a person using a bath or shower. For the purpose of this regulation the sheath of a silica glass sheathed element is regarded as part of the element.

601-12-02 Electric heating embedded in the floor shall be covered by an earthed metallic grid or have an earthed metallic sheath. This grid or sheath shall be connected to the local supplementary equipotential bonding specified in Regulation 601-04-02.

SECTION 602

SWIMMING POOLS

602-01   Scope

602-01-01   The particular requirements of this Section shall apply to basins of swimming pools and paddling pools and their surrounding zones where the risk of electric shock is increased by a reduction in body resistance and contact of the body with earth potential.

Special requirements may be necessary for swimming pools for medical use.

602-02   Assessment of general characteristics

Zone A   is the interior of the basin, chute or flume and includes the portions of essential apertures in its walls and floor which are accessible to persons in the basin.

Zone B is limited by :

> the vertical plane 2 m from the rim of the basin, and
>
> by the floor or surface expected to be occupied by persons, and
>
> by the horizontal plane 2.5 m above that floor or surface except where the basin is above ground, 2.5 m above the level of the rim of the basin.

Where the building containing the swimming pool contains diving boards, spring boards, starting blocks or a chute, Zone B includes also the Zone limited by - the vertical plane spaced 1.5 m from the periphery of diving boards, spring boards and starting blocks, and within that Zone, by the horizontal plane 2.5 m above the highest surface expected to be occupied by persons, or to the ceiling or roof if they exist.

Zone C   is limited by :

> the vertical plane circumscribing Zone B, and the parallel vertical plane 1.5 m external to Zone B, and
>
> by the floor or surface expected to be occupied by persons and the horizontal plane 2.5 m above that floor or surface.

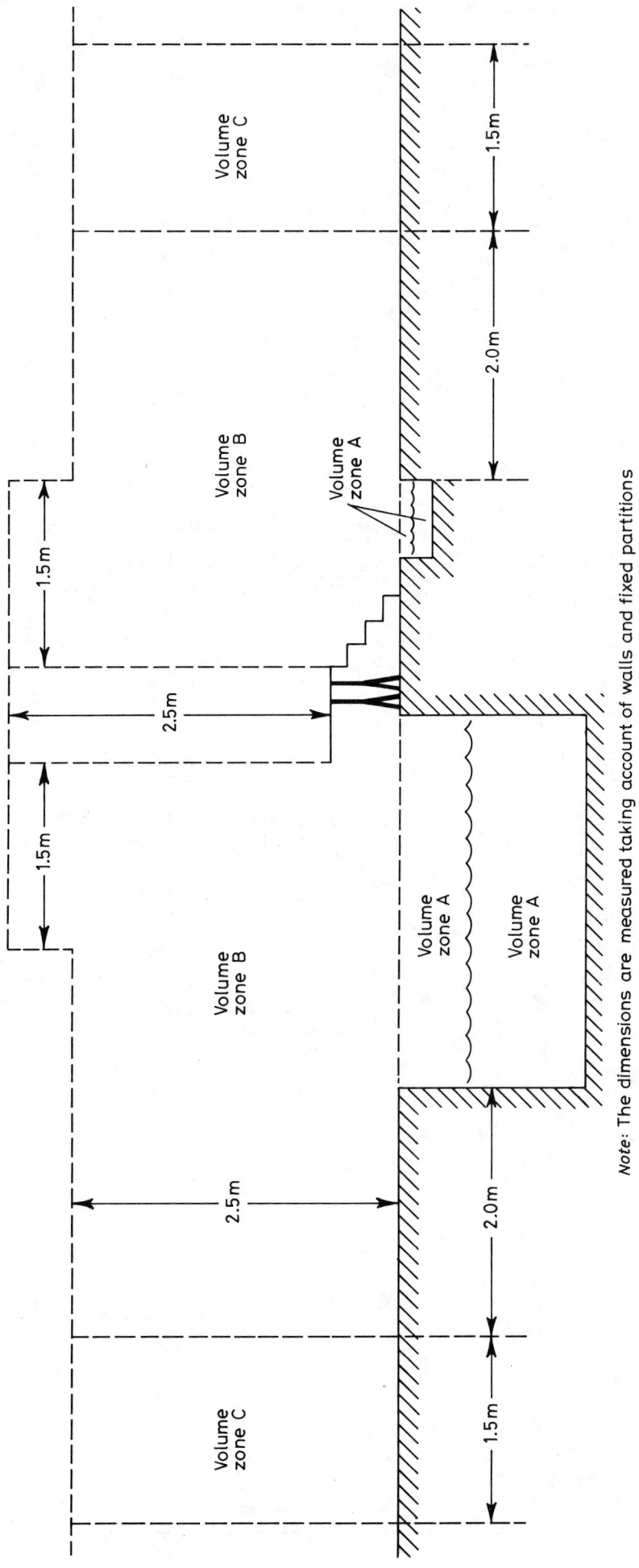

Figure 602A - Zone dimensions of swimming pools and paddling pools.

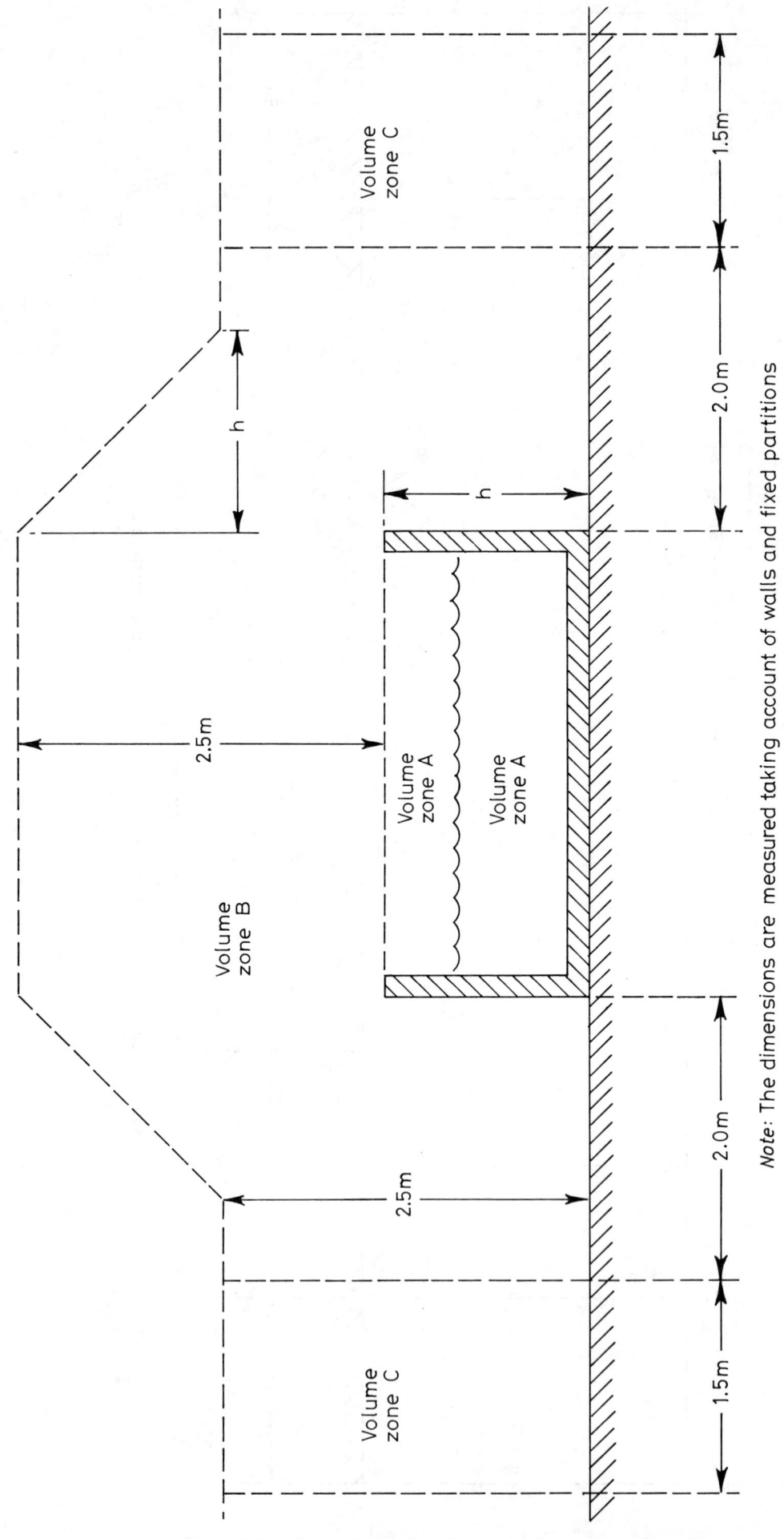

Figure 602B - Zone dimensions for basin above ground level.

Protection for safety

602-03 Protection against electric shock

602-03-01 Where SELV is used, irrespective of the nominal voltage, protection against direct contact shall be provided by:

(i) barriers or enclosures affording at least the degree of protection IP2X, or

(ii) insulation capable of withstanding a test voltage of 500 V a.c., r.m.s. for 1 minute.

602-03-02 Local supplementary equipotential bonding shall be provided connecting all extraneous-conductive--parts in Zone A, B and C together, with the protective conductors of all exposed-conductive-parts situated in these Zones. This requirement is not to be applied to equipment supplied by SELV circuits. An equipotential bonded grid shall be provided in all solid floors in Zones B and C.

602-04 Application of protective measures against electric shock

602-04-01 In Zones A and B, only the protective measure against electric shock by SELV (Regulation 411-02) at a nominal voltage not exceeding 12 V shall be used, the safety source being installed outside the Zones A, B and C, except that:

(i) where floodlights are installed, each floodlight shall be supplied from its own transformer (or an individual secondary winding of a multi-secondary transformer), having an open circuit voltage not exceeding 18 V

(ii) automatic disconnection of supply by means of a residual current device having the characteristics specified in Regulation 412-06-02 may be used to protect socket-outlets installed in accordance with Regulations 602- 07-01 or 602-07-02(iii).

602-04-02 The following protective measures shall not be used in any Zone:

(i) protection by means of obstacles (Regulation 412-04)

(ii) protection by means of placing out of reach (Regulation 412-05)

(iii) protection by means of a non-conducting location (Regulation 413-04)

(iv) protection by means of earth-free local equipotential bonding (Regulation 413-05).

Selection and erection of equipment

602-05 Degree of protection of enclosures

602-05-01 Equipment shall have the following minimum degrees of protection in accordance with BS 5490:

(i) in Zone A - IPX8

(ii) for swimming pools where water jets are likely to be used for cleaning purposes - IPX5

(iii) for swimming pools where water jets are not likely to be used for cleaning purposes.

        Zone B - IPX4
        Zone C - indoor pools IPX2
                     outdoor pools IPX4

602-06 Wiring systems

602-06-01 A surface wiring system shall not employ the use of metallic conduit or metallic trunking or an exposed metallic cable sheath or an exposed earthing or bonding conductor.

602-06-02 Zones A and B shall contain only wiring necessary to supply appliances situated in those Zones.

602-06-03 Junction boxes shall not be installed in Zones A and B.

602-07   Switchgear, controlgear and accessories

602-07-01   In Zones A and B, switchgear, controlgear and accessories shall not be installed except for swimming pools where it is not possible to locate socket-outlets outside Zone B, socket-outlets complying with BS 4343 may be installed outside arm's reach (i.e. 1.25 m) from the Zone A border and at least 0.3 m above the floor, only if they are protected by a residual current device complying with the relevant British Standard and having the characteristics specified in Regulation 412-06-02.

602-07-02   In Zone C, a socket-outlet, switch or accessory is permitted only if it is:

(i)   protected individually by electrical separation, (Regulation 413-06), or

(ii)   protected by SELV (Regulation 411-02), or

(iii) protected by a residual current device complying with the appropriate British Standard and having the characteristics specified in Regulation 412-06-02, or

(iv)   a shaver socket complying with BS 3535.

This requirement does not apply to the insulating cords of cord operated switches complying with BS 3676.

602-08   Other equipment

602-08-01   Socket-outlets shall comply with BS 4343.

602-08-02   In Zones A and B, only fixed equipment specifically intended for use in swimming pools shall be installed.

602-08-03   In Zone C, equipment shall be protected by one of the following:

(i)   individually by electrical separation (Regulation 413-06)

(ii)   SELV (Regulation 411-02)

(iii) a residual current device having the characteristics specified in Regulation 412-06-02.

This requirement does not apply to instantaneous water heaters complying with the relevant Section of BS 3456.

602-08-04   An electric heating unit embedded in the floor in Zones B and C shall incorporate a metallic sheath connected to the local supplementary equipotential bonding and shall be covered by the metallic grid required by Regulation 602-03-02.

602-08-05   In Zone B, only water heaters are permitted excepting that other equipment supplied by SELV at a nominal voltage not exceeding 12 V may be installed.

SECTION 603

HOT AIR SAUNAS

603-01 Scope

603-01-01 The particular requirements of this section shall apply to locations in which hot air sauna heating equipment according to IEC Publication 335-2-53 (Safety of Household and Similar Electrical Appliances Part 2: Particular requirements for electric sauna heating appliances) is installed.

603-02 Classification of temperature zones

603-02-01 The assessment of the general characteristics of the location shall take due consideration of the classification of the four temperature zones which are illustrated in Figure 603A.

Protection against electric shock

603-03 Protection against both direct and indirect contact

603-03-01 Where SELV is used, irrespective of the nominal voltage, protection against direct contact shall be provided by one or more of the following:

(i) insulation (Regulation 412-02), capable of withstanding a test voltage of 500 V a.c., r.m.s. for 60 s

(ii) barriers or enclosures (Regulation 412-03), affording at least the degree of protection IP 24 (see BS 5490).

Application of protective measures against electric shock

603-04 Protection against direct contact

603-04-01 The following protective measures against direct contact shall not be used:

(i) protection by means of obstacles (Regulation 412-04)

(ii) protection by placing out of reach (Regulation 412-05).

603-05 Protection against indirect contact

603-05-01 The following protective measures against indirect contact shall not be used:

(i) protection by non-conducting location (Regulation 413- 04)

(ii) protection by means of earth-free local equipotential bonding (Regulation 413-05).

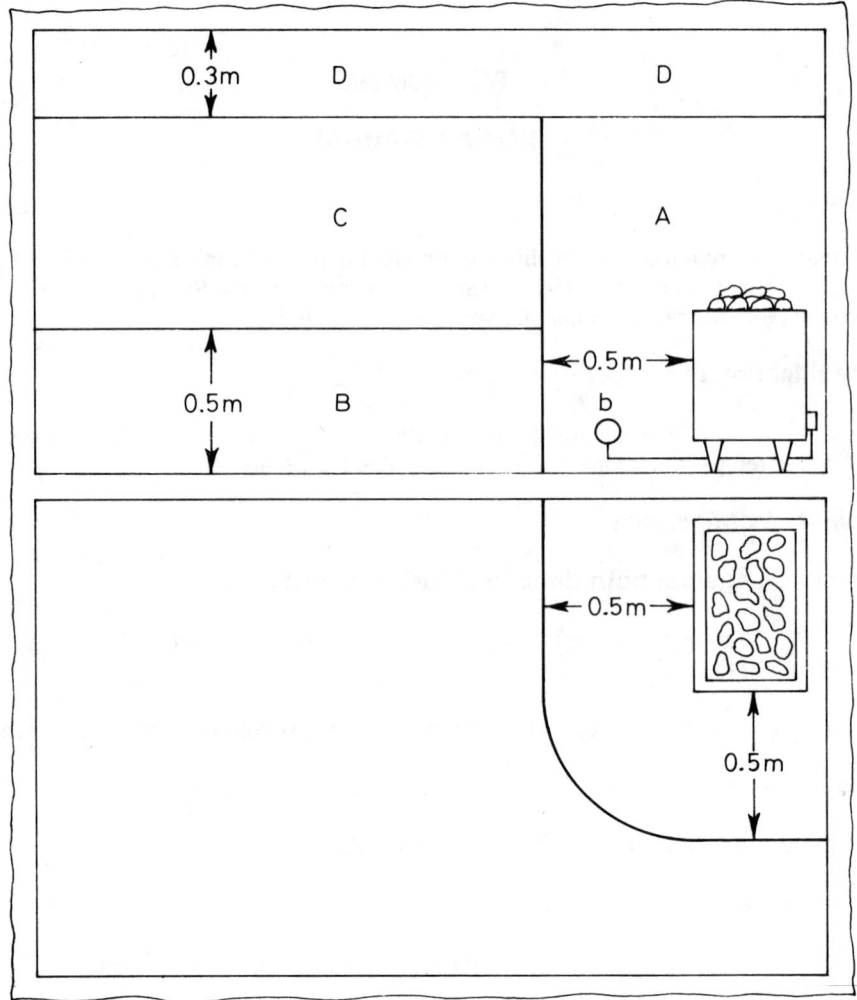

Figure 603A

Selection and erection of equipment

603-06   Common Rules

603-06-01   All equipment shall have at least the degree of protection IP 24 to BS 5490.

603-06-02   In temperature Zone A only the sauna heater and equipment directly associated with it shall be installed.

In temperature Zone B there is no special requirement concerning heat resistance of equipment.

In temperature Zone C equipment shall be suitable for an ambient temperature of 125 $^{\circ}$C.

In temperature Zone D only luminaires and their associated wiring, and control devices for the sauna heater and their associated wiring shall be installed. The equipment shall be suitable for an ambient temperature of 125 $^{\circ}$C.

603-07   Wiring Systems

603-07-01   Only flexible cords complying with BS 6141 having 150 $^{\circ}$C rubber insulation shall be used and shall be mechanically protected with material which complies with Regulation 413-03-01.

603-08   Switchgear, controlgear and accessories

603-08-01   Switchgear not built into the sauna heater, other than a thermostat and a thermal cut-out shall be installed outside the hot air sauna.

603-08-02   Except as permitted in Regulations 603-06-02 and 603-08-01 accessories shall not be installed within the hot air sauna.

603-09   Other fixed equipment

603-09-01   Luminaires shall be so mounted as to prevent overheating.

SECTION 604

CONSTRUCTION SITE INSTALLATIONS

604-01 Scope

604-01-01 The particular requirements of this Section shall apply to temporary installations providing supplies during the execution of the following works:

(i) new building construction

(ii) repair, alteration, extension or demolition of existing buildings

(iii) engineering construction

(iv) earthworks

(v) similar works.

The requirements of this Section shall not apply to installations in construction site offices, cloakrooms, meeting rooms, canteens, restaurants, dormitories, toilets, etc., where the general requirements of the Regulations shall apply.

Construction site fixed installations are limited to the assembly of the main switchgear and principal protective devices.

An installation on the load side is considered a movable installation except for parts which are designed according to Chapter 52 of these Regulations.

Assessment and general characteristics

604-02 Supplies

604-02-01 Equipment shall be identified with and be compatible with the particular supply from which it is energised and shall contain only components connected to one and the same installation, except for control or signalling circuits and input from standby supplies.

604-02-02 The following voltages shall not be exceeded:

(i) 25 V, 1 phase, SELV - Portable hand lamps in confined and damp locations

(ii) 50 V, 1 phase, centre point earthed -Portable hand lamps in confined and damp locations

(iii) 110 V, 1 phase, centre point earthed - reduced low voltage system
    Portable hand lamps for general use
    Portable handheld tools and local lighting up to 2 kW

(iv) 110 V, 3 phase, star point earthed - reduced low voltage system
    Small mobile plant, up to 3.75 kW

(v) 240 V, 1 phase - fixed floodlighting
(vi) 415 V, 3 phase - fixed and movable equipment, above 3.75 kW.

This requirement shall not be deemed to preclude the use of a high voltage supply for large equipment where this is necessary for functional reasons.

Protection for safety

604-03 Protection against indirect contact

604-03-01 Where an alternative system is available an IT system shall not be used. Where an IT system is used, permanent earth fault monitoring shall be provided.

604-03-02 Where protection against indirect contact is provided by the measure of earthed equipotential bonding and automatic disconnection of supply, (Regulations 413-02 and 471-08 as appropriate to the type of earthing system), then Regulations 604-04 to 604-08 shall apply.

604-04 TN system

604-04-01 In Regulation 413-02-08, Table 41A is replaced by Table 604A.

TABLE 604A

Maximum disconnection times for TN systems

(see Regulation 604-04-02)

U_o (volts)	t* (seconds)
120	0.35
220 to 277	0.2
400, 480	0.05
580	0.02

U_o Nominal voltage to Earth

*   If such a disconnecting time cannot be guaranteed it may be necessary to take other protective measures, such as supplementary equipotential bonding.

604-04-02 Regulation 413-02-09 is replaced by:

Except for a reduced low voltage system, (see Regulation 471-15) the maximum disconnection times stated in Table 604A shall apply to circuits supplying movable installations or equipment, either directly or through socket-outlets.

604-04-03 Regulation 413-02-10 is replaced by:

Where a fuse is used to satisfy the requirements of Regulation 604-04-02, maximum values of earth fault loop impedance (Z_s) corresponding to a disconnection time of 0.2 s are stated in Table 604B1 for a nominal voltage to Earth (U_o) of 240 V. For types and rated currents of general purpose (gG) fuses other than those mentioned in Table 604B1, and for motor circuit fuses (gM), reference should be made to the appropriate British Standard, to determine the value of I_a for compliance with Regulation 604-04-02.

TABLE 604B1

Maximum earth fault loop impedance (Z_s) for fuses, for 0.2 s disconnection time with U_o 240 V (see Regulation 604-04-03)

(a) General purpose (gG) fuses to BS 88 Parts 2 and 6

Rating (amperes)	6	10	16	20	25	32	40	50
Z_s (ohms)	7.74	4.71	2.53	1.60	1.33	0.92	0.71	0.53

(b) Fuses to BS 1361

Rating (amperes)	5	15	20	30	45
Z_s (ohms)	9.60	3.00	1.55	1.00	0.51

(c) Fuse to BS 3036

Rating (amperes)	5	15	20	30	45
Z_s (ohms)	7.50	1.92	1.33	0.80	0.41

(d) Fuse to BS 1362

Rating (amperes)	13
Z_s (ohms)	2.14

604-04-04   Regulation 413-02-11 is replaced by:

Where a circuit-breaker is used to satisfy the requirements of Regulation 604-04-02, the maximum value of earth fault loop impedance (Z_s) shall be determined by the formula of Regulation 413-02-08. Alternatively, for a nominal voltage to earth of 240 V and a disconnection time of 0.2 s, the values specified in Table 604B2 for the types and ratings of circuit-breaker listed may be used instead of calculation.

TABLE 604B2

Maximum earth fault loop impedance (Z_s) for miniature circuit-breakers, for disconnection times of both 0.2 s with U_o 240 V (see Regulation 604-04-04) and 5 s (see Regulations 413-02-12 and 604-04-06)

(e) Type 1 miniature circuit-breakers to BS 3871

Rating (amperes)	5	6	10	15	16	20	30	32	40	45	50	63	I_n
Z_s (ohms)	12.0	10.0	6.0	4.0	3.75	3.0	2.0	1.88	1.5	1.33	1.20	0.95	$240/(4I_n)$

(f) Type 2 miniature circuit-breakers to BS 3871

Rating (amperes)	5	6	10	15	16	20	30	32	40	45	50	63	I_n
Z_s (ohms)	6.86	5.71	3.43	2.29	2.14	1.71	1.14	1.07	0.86	0.76	0.69	0.54	$240/(7I_n)$

(g) Type B miniature circuit-breakers to BS 3871

Rating (amperes)	6	10	16	20	32	40	45	50	63	I_n
Z_s (ohms)	8.0	4.80	3.00	2.40	1.50	1.20	1.07	0.96	0.76	$240/(5I_n)$

(h) Type 3 and Type C miniature circuit-breakers to BS 3871

Rating (amperes)	5	6	10	15	16	20	30	32	40	45	50	63	I_n
Z_s (ohms)	4.80	4.0	2.40	1.60	1.50	1.20	0.80	0.75	0.60	0.53	0.48	0.381	$240/(10I_n)$

604-04-05   Regulation 413-02-12 is not applicable. Table 41C is not applicable.

604-04-06   Regulation 413-02-13 is replaced by:

A maximum disconnection time of 5 s shall apply to the fixed installation and to a reduced low voltage system (see Regulation 471-15).

For a nominal voltage to Earth (U_o) of 240 V r.m.s. a.c. and for types and rated currents of overcurrent protective devices in common use, maximum values of earth fault loop impedance (Z_s) corresponding to a disconnection time of 5 s are stated in Regulation 413-02-14.

604-04-07   Regulation 413-02-15 is replaced by:

If the conditions of Regulations 604-04-01 and 604-04-06 cannot be fulfilled by using overcurrent protective devices, then protection shall be provided by means of a residual current device.

604-04-08   In Regulation 413-02-16, the formula is replaced by:

$$Z_s \, I_{\Delta n} \leq 25 \text{ V}$$

604-05 TT system

604-05-01 In Regulation 413-02-20, the formula is replaced by:

$$R_a \; I_a \leqslant 25 \text{ V}$$

604-06 IT system

604-06-01 In Regulation 413-02-23, the formula is replaced by:

$$R_b \; I_d \leqslant 25 \text{ V}$$

604-06-02 In Regulation 413-02-26, Table 41E is replaced by Table 604E and the definition of I_a is replaced by the following:

I_a is the current which disconnects the circuit within the time specified in Table 604E when applicable, or within 5 s for other circuits when this time is allowed (see Regulation 604-04-06).

TABLE 604E

Maximum disconnecting time in IT sytems (Second fault)

U_0/ U (volts)	Neutral not distributed t (seconds)*	Neutral distributed t (seconds)*
120-240	0.4	1
220/380 to 277/480	0.2	0.5
400/690	0.06	0.2
580/1000	0.02	0.08

U_0 Voltage between phase and neutral

U Voltage between phases

* If such disconnecting time cannot be guaranteed, it may be neccessary to take other protective measures, such as a supplementary equipotential bonding.

604-07 Supplementary equipotential bonding

604-07-01 In Regulation 413-02-28, the formula is replaced by:

$$R \leqslant \frac{25}{I_a}$$

604-08 Application of protective measures

604-08-01 Regulation 471-08-02 is not applicable.

604-08-02 Regulation 471-08-03 is not applicable.

604-08-03 Regulations 471-08-04 and 471-08-05 are not applicable. Every socket-outlet shall be protected by one or more of the following:

(i) automatic disconnection and reduced low voltage system (Regulation 471-15)

(ii) a residual current device complying with the appropriate British Standard and having the characteristics specified in Regulation 412-06-02(ii)

(iii) SELV (Regulations 411-02 and 471-02)

(iv) electrical separation (Regulations 413-06 and 471-12).

604-08-04 Where socket-outlets are protected in accordance with Regulation 604-08-03(iv), each socket-outlet shall be supplied from a separate transformer.

Selection and erection of equipment

604-09 General

604-09-01 Every assembly for the distribution of electricity on construction and demolition sites shall comply with the requirements of BS 4363 and BS 5486: Part 4.

604-09-02 Except for assemblies covered by Regulation 604-09-01 equipment shall have a degree of protection of at least IP 44. (BS 5490)

604-10 Wiring systems

604-10-01 Every wiring system shall be so arranged that no strain is placed on the terminations of conductors unless such terminations are designed for this purpose.

604-10-02 Cable shall not be run across a site road or a walkway unless adequate protection of the cable against mechanical damage is provided.

604-11 Isolation and switching devices

604-11-01 At the origin of each installation, an assembly comprising the main controlgear and principal protective devices shall be provided.

604-11-02 Each supply assembly and each distribution assembly shall incorporate means for isolating and switching the incoming supply.

604-11-03 A means of emergency switching shall be provided on the supply to all equipment from which it may be necessary to disconnect all live conductors in order to remove a hazard.

604-11-04 Every isolating device for isolation of the incoming supply shall be suitable for securing in the 'off' position.

604-11-05 Every circuit supplying equipment shall be fed from a distribution assembly comprising the following:

(i) overcurrent protective devices

(ii) devices affording protection against indirect contact

(iii) socket-outlets, if required.

604-11-06 Safety and standby supplies shall be connected by means of devices arranged to prevent interconnection of different supplies.

604-12 Plugs and socket-outlets

604-12-01 Every socket-outlet shall be incorporated as part of an assembly complying with Regulation 604-09-01.

604-12-02 Every plug and socket-outlet shall comply with BS 4343.

604-12-03 Luminaire supporting couplers shall not be used.

604-13 Cable couplers

604-13-01 Every cable coupler shall comply with BS 4343.

SECTION 605

AGRICULTURAL AND HORTICULTURAL PREMISES

605-01 Scope

605-01-01 The particular requirements of this Section shall apply to all parts of fixed installations of agricultural and horticultural premises outdoors and indoors, and to locations where livestock is kept. Where the above premises include dwellings intended solely for human habitation the dwellings are excluded from the scope of this Section.

Protection for safety

605-02 Protection against both direct and indirect contact

605-02-01 Where protection by the use of SELV (Regulations 411-02 and 471-02) is used in situations accessible to livestock in and around agricultural buildings, the upper limit of nominal voltage specified in Regulation 411-02-01(i) does not apply and shall be reduced as appropriate to the type of livestock.

605-02-02 Where SELV is used, irrespective of the nominal voltage, protection against direct contact shall be provided by one or more of the following:

(i) barriers or enclosures (Regulation 412-03), affording at least the degree of protection IP2X (see BS 5490)

(ii) insulation (Regulation 412-02) capable of withstanding a test voltage of 500 V a.c., r.m.s. for 60 s.

605-03 Protection against direct contact

605-03-01 Every circuit supplying a socket-outlet shall be protected by a residual current device complying with the appropriate British Standard and having the characteristics specified in Regulation 412-06-02(ii).

605-04 Protection against indirect contact

605-04-01 In locations in which livestock is kept and, where protection against indirect contact is provided by the measure of earthed equipotential bonding and automatic disconnection of supply, (Regulations 413-02 and 471-08 as appropriate to the type of earthing system) then Regulations 605-05 to 605-09 shall apply.

605-05 TN system

605-05-01 In Regulation 413-02-08, Table 41A is replaced by Table 605A.

TABLE 605A

Maximum disconnecting time for TN systems (see Regulation 605-05-02)

U_o (volts)	t (seconds)*
120	0.35
220 to 277	0.2
400, 480	0.05
580	0.02

U_o Nominal voltage to Earth

* If such disconnecting time cannot be guaranteed, it may be necessary to take other protective measures, such as a supplementary equipotential bonding.

605-05-02 In Regulation 413-02-09, Table 41A is replaced by Table 605A.

605-05-03 Regulation 413-02-10 is replaced by:

Where a fuse is used to satisfy the requirements of Regulation 605-05-02, maximum values of earth fault loop impedance (Z_s) corresponding to a disconnection time of 0.2 s are stated in Table 605B1 for a nominal voltage to Earth (U_o) of 240 V. For types and rated currents of general purpose (gG) fuses other than those mentioned in Table 605B1, and for motor circuit fuses (gM), reference should be made to the appropriate British Standard, to determine the value of I_a for compliance with Regulation 605-05-02.

TABLE 605B1

Maximum earth fault loop impedance (Z_s) for fuses, for 0.2 s disconnection time with U_o 240 V (see Regulation 605-05-03)

(a) General purpose (gG) fuses to BS 88 Parts 2 and 6

Rating (amperes)	6	10	16	20	25	32	40	50
Z_s (ohms)	7.74	4.71	2.53	1.60	1.33	0.92	0.71	0.53

(b) Fuses to BS 1361

Rating (amperes)	5	15	20	30	45
Z_s (ohms)	9.60	3.0	1.55	1.0	0.51

(c) Fuses to BS 3036

Rating (amperes)	5	15	20	30	45
Z_s (ohms)	7.50	1.92	1.33	0.80	0.41

(d) Fuse to BS 1362

Rating (amperes)	13
Z_s (ohms)	2.14

605-05-04 Regulation 413-02-11 is replaced by:

Where a circuit-breaker is used to satisfy the requirements of Regulation 605-05-02, the maximum value of earth fault loop impedance (Z_s) shall be determined by the formula of Regulation 413-02-08. Alternatively, for a nominal voltage to earth of 240 V and a disconnection time of 0.2 s, the values specified in Table 605B2 for the types and ratings of circuit-breaker listed may be used instead of calculation.

TABLE 605B2

Maximum earth fault loop impedance (Z_s) for miniature circuit-breakers, for disconnection times of both 0.2 s with U_o 240 V (see Regulation 605-05-04) and 5 s (see Regulations 413-02-12 and 605-05-06)

(e) Type 1 miniature circuit-breakers to BS 3871

Rating (amperes)	5	6	10	15	16	20	30	32	40	45	50	63	I_n
Z_s (ohms)	1.20	10.0	6.00	4.00	3.75	3.00	2.00	1.88	1.50	1.33	1.20	0.95	$240/(4I_n)$

(f) Type 2 miniature circuit-breakers to BS 3871

Rating (amperes)	5	6	10	15	16	20	30	32	40	45	50	63	I_n
Z_s (ohms)	6.86	5.71	3.43	2.29	2.14	1.71	1.14	1.07	0.86	0.76	0.69	0.54	$240/(7I_n)$

(g) Type B miniature circuit-breakers to BS 3871

Rating (amperes)	6	10	16	20	32	40	45	50	63	I_n
Z_s (ohms)	8.00	4.80	3.00	2.40	1.50	1.20	1.07	0.96	0.76	$240/(5I_n)$

(h) Type 3 and Type C miniature circuit-breakers to BS 3871

Rating (amperes)	5	6	10	15	16	20	30	32	40	45	50	63	I_n
Z_s (ohms)	4.80		2.40		1.50		0.80		0.60		0.48		$240/(10I_n)$
		4.00		1.60		1.20		0.75		0.53		0.38	

605-05-05 Regulation 413-02-12 is not applicable.

Table 41C is not applicable.

605-05-06 Regulation 413-02-13 is replaced by:

For a distribution circuit a disconnection time not exceeding 5 s is permitted.

For a final circuit supplying only stationary equipment and for a final circuit for which the requirement of Regulation 413-02-09 does not apply a disconnection time not exceeding 5 s is permitted. Where the disconnection time for such a final circuit exceeds that required by Table 605A and another final circuit requiring a disconnection time according to Table 605A is connected to the same distribution board or distribution circuit, one of the following conditions shall be fulfilled:

(i) the impedance of the protective conductor between the distribution board and the point at which the protective conductor is connected to the main equipotential bonding shall not exceed 25 Z_s/U_o ohms (where Z_s is the earth fault loop impedance corresponding to a disconnection time of 5 s)

(ii) there shall be equipotential bonding at the distribution board which involves the same types of extraneous-conductive-parts as the main equipotential bonding according to Regulation 413-02-02 and is sized in accordance with Regulation 547-02-01.

605-05-07 In Regulation 413-02-14, Regulation 413-02-13 is replaced by Regulation 605-05-06.

605-05-08 In Regulation 413-02-15, Regulations 413-02-08 to 413-02-14 are replaced by Regulations 605-05-01, 605-05-02, 605-05-03, 605-05-04, 605-05-06 and 605-05-07.

605-05-09 In Regulation 413-02-16, the formula is replaced by:

$$Z_s \, I_{\Delta n} \leq 25 \text{ V}$$

605-06 Installations which are part of a TT system

605-06-01 In Regulation 413-02-20, the formula is replaced by:

$$R_a \, I_a \leq 25 \text{ V}$$

605-07 Installations which are part of an IT system

605-07-01 In Regulation 413-02-23, the formula is replaced by:

$$R_b \, I_d \leq 25 \text{ V}$$

605-07-02 In Regulation 413-02-26, Table 41E is replaced by Table 605E and the definition of I_a is replaced by the following:

I_a is the current which disconnects the circuit within the time t specified in Table 605E when applicable, or within 5 s for other circuits when this time is allowed (see Regulation 605-05-06).

TABLE 605E

Maximum disconnection time in IT systems (Second fault) (see Regulation 605-07-02)

Installation nominal voltage	Disconnection time t (seconds)*	
U_0/U (volts)	Neutral not distributed	Neutral distributed
120-240	0.4	1
220/380 to 277/480	0.2	0.5
400/690	0.06	0.2
580/1000	0.02	0.08

U_0 Voltage between phase and neutral

U Voltage between phases

* If such disconnecting time cannot be guaranteed, it may be necessary to take other protective measures, such as a supplementary equipotential bonding.

605-08 Supplementary equipotential bonding

605-08-01 In Regulation 413-02-28, the formula is replaced by:

$$R \leq \frac{25}{I_a}$$

605-08-02 In locations where livestock is kept supplementary equipotential bonding shall connect all those exposed-conductive-parts and extraneous-conductive-parts which can be touched by livestock. Extraneous-conductive-parts include non-insulating floors.

605-08-03 Where a metallic grid is laid in the floor for supplementary bonding it shall be connected to the protective conductors of the installation.

Application of protective measures against indirect contact

605-09 Protection by earthed equipotential bonding and automatic disconnection of supply

605-09-01 Regulation 471-08-02 is replaced by the following:

For an installation which is part of a TN system, the limiting values of earth fault loop impedance and of circuit protective conductor impedance specified by Regulations 605-05-01, 605-05-03, 605-05-04, 605-05-06 and 605- 05-07 are applicable only where the exposed-conductive-parts of the equipment concerned and any extraneous-conductive-parts are situated within the earthed equipotential zone created by the main equipotential bonding (see also Regulation 413-02-02).

Where the disconnection times specified by Regulation 605-05-01 cannot be met by the use of an overcurrent protective device, Regulation 605-05-08 shall be applied.

605-09-02 In Regulation 471-08-03, Table 41A is replaced by Table 605A.

605-09-03 Regulation 471-08-06 is not applicable and Regulation 605-03-01 applies.

605-10 Protection against fire and harmful thermal effects

605-10-01 For the purpose of protection against fire, a residual current device, having a rated residual operating current not exceeding 0.5 A, shall be installed for the supply to equipment other than that essential to the welfare of livestock.

605-10-02 Heating appliances shall be fixed so as to maintain an appropriate distance from livestock and combustible material, to minimise any risks of burns to livestock and of fire. For radiant heaters the clearance shall be not less than 0.5 m or such other clearance as recommended by the manufacturer.

Selection and erection of equipment

605-11 External influences

605-11-01 In situations accessible to livestock in and around agricultural buildings, electrical equipment shall so far as is practicable be of Class II construction, or constructed of or protected by suitable insulating materials, and shall be of a type appropriate to the other external influences likely to occur including dust and water ingress, corrosion, mechanical stresses, flora, fauna, solar radiation and wind.

605-12 Wiring systems

605-12-01 In locations where livestock is kept fixed wiring systems shall be inaccessible to livestock.

605-12-02 Cables liable to attack by vermin shall be of a suitable type or be suitably protected.

Switchgear and controlgear

605-13 Devices for isolation and switching

605-13-01 Each device for emergency switching including emergency stopping shall be installed where it is inaccessible to livestock and will not be impeded by livestock, due account being taken of conditions likely to arise in the event of panic by livestock.

Other equipment

605-14 Electric fence controllers

605-14-01 Mains operated electric fence controllers shall comply with BS 2632 or BS 6369.

605-14-02 Every mains-operated electric fence controller shall be so installed that, so far as is reasonably practicable, it is free from risk of mechanical damage or unauthorised interference.

605-14-03 A mains-operated fence controller shall not be fixed to any supporting pole of an overhead power or telecommunication line; provided that, where a low voltage supply to an electric fence controller is carried by an insulated overhead line from a distribution board, the controller may be fixed to the pole carrying the supply.

605-14-04 Every earth electrode which is connected to the earthing terminal of an electric fence controller shall be separate from the earthing system of any other circuit and shall be situated outside the resistance area of any electrode used for protective earthing.

605-14-05 Not more than one controller shall be connected to each electric fence or similar system of conductors.

605-14-06 Every electric fence or similar system of conductors and the associated controller shall be so installed that it is not liable to come into contact with any other equipment or conductor.

SECTION 606

RESTRICTIVE CONDUCTIVE LOCATIONS

606-01 Scope

606-01-01 The particular requirements of this Section shall apply to installations within or intended to supply equipment or appliances to be used within a Restrictive Conductive Location. They do not apply to any location in which freedom of movement is not physically constrained.

Protection for safety

606-02 Protection against direct and indirect contact

606-02-01 Where protection by the use of SELV or functional extra-low voltage (Regulations 411- 02 to 411-03 and 471-02) is used the voltage shall not exceed 25 V a.c., r.m.s. or 60 V ripple free d.c. and, regardless of the voltage, protection against direct contact shall be provided by:

(i) a barrier or enclosure affording at least the degree of protection IP2X (see BS 5490), or

(ii) insulation capable of withstanding a test voltage of 500 V a.c. r.m.s. for 60 s.

606-03 Protection against direct contact

606-03-01 Protection by the following means is not permitted:

(i) obstacles (Regulations 412-04 and 471-06)

(ii) placing out of reach (Regulations 412-05 and 471-07).

606-04 Protection against indirect contact

606-04-01 Protection against indirect contact shall be provided by one of the following:

(i) SELV (Regulation 606-02)

(ii) FELV (Regulation 606-02), in which case a supplementary equipotential bonding conductor (Regulations 413-02-27 and 28) shall be provided and be connected to the exposed-conductive-parts, the conductive parts of the location and the functional earth terminal of the equipment or socket

(iii) automatic disconnection (Regulations 413-02-01 to 413-02- 26 and 471-08) in which case a supplementary equipotential bonding conductor (Regulations 413-02-27 and 28) shall be provided and be connected to the exposed conductive parts of the fixed equipment and the conductive parts of the location

(iv) electrical separation (Regulation 413-06) in which case only one socket or piece of equipment shall be connected to each secondary winding of the isolating transformer

(v) the use of Class II equipment adequately protected to an IP code (see BS 5490 and Regulation 512-06) in which case the circuit shall be further protected by a residual current device having the characteristics specified in Regulation 412-06-02.

606-04-02 A supply to or a socket intended to supply a hand lamp shall be protected by SELV (Regulations 606-02 and 606-04-01(i)).

606-04-03 A supply to or a socket intended to supply equipment requiring a functional earth shall be protected by FELV (Regulations 606-02 and 606- 04-01(ii)).

606-04-04 A supply to or a socket intended to supply a hand held tool shall be protected by SELV (Regulations 606-02 and 606-04-01(i)) or electrical separation (Regulation 606-04-01(iv)).

606-04-05 A supply to fixed equipment shall be protected by one of the methods listed in Regulation 606-04-01.

606-04-06 Every safety source and isolating source, other than those specified in Regulation 411-02-02 (iii), shall be situated outside the restrictive conductive location, unless it is part of the fixed installation within a permanent restrictive conductive location as provided by Regulation 606-04-01.

SECTION 607

EARTHING REQUIREMENTS FOR THE INSTALLATION OF EQUIPMENT HAVING HIGH EARTH LEAKAGE CURRENTS

607-01 Scope

607-01-01 The particular requirements of this Section shall apply to every installation supplying equipment having a high earth leakage current (usually exceeding 3.5 mA), including information technology equipment to BS 7002 and industrial control equipment where values of earth leakage current in normal service permitted by British Standards necessitate special precautions being taken in the installation of the equipment.

607-02 General

607-02-01 Where a low-noise earth is specified for a particular item of equipment, the requirement of Sections 545 and 546 shall apply.

607-02-02 Except as required by Regulation 607-02-06 no special precaution is necessary for equipment complying with BS 7002, and having leakage current not exceeding 3.5 mA.

607-02-03 Where more than one item of stationary equipment having an earth leakage current exceeding 3.5 mA in normal service is to be supplied from an installation incorporating a residual current device, it shall be verified that the total leakage current does not exceed 25% of the nominal tripping current of the residual current device (see also Regulation 531-02-03).

Where compliance with this Regulation cannot be otherwise achieved the items of equipment shall be supplied through a double-wound transformer or equivalent device as described in Item (vi) of Regulation 607-02-07.

607-02-04 An item of stationary equipment having an earth leakage current exceeding 3.5 mA but not exceeding 10 mA in normal service shall either be permanently connected to the fixed wiring of the installation without the use of a plug and socket-outlet or shall be connected by means of a plug and socket-outlet complying with BS 4343.

607-02-05 An item of stationary equipment having an earth leakage current exceeding 10 mA in normal service shall preferably be permanently connected to the fixed wiring of the installation. Alternatively, the equipment may be connected by means of a plug and socket-outlet complying with BS 4343 provided that the protective conductor of the associated flexible cable is supplemented by a separate contact and protective conductor having a cross-sectional area not less than 4 mm^2 or the flexible cable complies with 607-02-07(iii) with the second protective conductor connected via a separate contact within the plug. The permanent connection to the fixed wiring may be by means of a flexible cable.

607-02-06 For a final circuit supplying a number of socket-outlets in a location intended to accommodate several items of equipment, where it is known or is reasonably to be expected that the total earth leakage current in normal service will exceed 10 mA, the circuit shall be provided with a high integrity protective connection complying with one or more of the arrangements described in Items (i) to (vi) of Regulation 607-02-07. Alternatively a ring circuit may be used to supply a number of single socket-outlets. There shall be no spur from the ring and the supply ends of the protective conductor ring shall be separately connected at the distribution board. The minimum size of the protective earth conductor ring shall be 1.5 mm^2.

607-02-07 The fixed wiring of every final circuit intended to supply an item of stationary equipment having an earth leakage current exceeding 10 mA in normal service shall be provided with a high integrity protective connection complying with one or more of the arrangements described in Items (i) to (vi) below:

(i) a single protective conductor with a cross-sectional area of not less than 10 mm^2

(ii) separate duplicated protective conductors, having independent connections complying with Regulation 526-01, each having a cross-sectional area not less than 4 mm^2

(iii) duplicate protective conductors incorporated in a multicore cable together with the live conductors of the circuit, provided that the total cross-sectional area of all the conductors of the cable is not less than 10 mm^2. One of the protective conductors may be formed by a metallic armour, sheath or braid incorporated in the construction of the cable and complying with Regulation 543-02-05

(iv) duplicate protective conductors formed by metal conduit, trunking or ducting complying with Regulation 543-02-04, and by a conductor having a cross-sectional area not less than 2.5 mm^2 installed in the same enclosure and connected in parallel with it

(v) an earth monitoring device which, in the event of a discontinuity in the protective conductor, automatically disconnects the supply of the equipment in accordance with Regulations 413-02-01 to 413-02-26

(vi) connection of the equipment to the supply by means of a double-wound transformer or other unit in which the input and output circuits are electrically separated, the circuit protective conductor is connected to the exposed-conductive-parts of the equipment and to a point of the secondary winding of the transformer or equivalent device. The protective conductor(s) between the equipment and the transformer shall comply with one of the arrangements described in Items (i) to (iv) above.

Except where Regulation 607-02-05 applies each protective conductor mentioned in Item (i) to (iv) above shall comply with the requirements of Sections 413 and 543.

607-03 Requirements for TT system

607-03-01 Where items of stationary equipment having an earth leakage current exceeding 3.5 mA in normal service are to be supplied from an installation forming part of a TT system, it shall be verified that the product of the total earth leakage current (in amperes) and twice the resistance of the installation earth electrodes (in ohms) does not exceed 50.

Where compliance with this Regulation cannot be otherwise achieved the items of equipment shall be supplied through a double-wound transformer or equivalent device as described in Item (vi) of Regulation 607-02-07.

607-04 Requirements for IT system

607-04-01 Equipment having a high earth leakage current shall not be connected directly to an IT system.

Selection and erection of equipment

607-05 Safety requirement for low noise earthing arrangements

607-05-01 Each exposed-conductive-part of data processing equipment shall be connected to the main earthing terminal.

This requirement shall also apply to metallic enclosures of Class II and Class III equipment and to a functional extra-low voltage circuit when this is earthed for functional reasons.

An earth conductor which serves a functional purpose only is not required to comply with Section 543.

SECTION 608

DIVISION ONE

ELECTRICAL INSTALLATIONS IN CARAVANS AND MOTOR CARAVANS

608-01 Scope

608-01-01 The particular requirements of this section apply to the electrical installations of caravans and motor caravans at nominal voltages not exceeding 250 V single phase. They do not apply to those electrical circuits and equipment covered by the Road Vehicles Lighting Regulations 1989 nor to installations covered by BS 6765 Part 3.

The particular requirements of Section 601 apply also to such installations in caravans or motor caravans.

Protection against shock

608-02 Protection against direct contact

608-02-01 The following methods of protection shall not be used:

(i) protection by obstacles (Regulations 412-04 and 471-06)

(ii) protection by placing out of reach (Regulations 412-05 and 471-07).

608-03 Protection against indirect contact

608-03-01 The following methods of protection shall not be used:

(i) non-conducting location (Regulations 413-04 and 471-10)

(ii) earth free equipotential bonding (Regulations 413-05 and 471-11)

(iii) electrical separation (Regulations 413-06 and 471-12).

608-03-02 Where protection by automatic disconnection of supply is used (Regulation 413-02 and 471-08) a double pole residual current device shall be provided complying with the relevant British Standard and having the characteristics specified in Regulation 412-06-02 and the wiring system shall include a circuit protective conductor which shall be connected to:

(i) the protective contact of the inlet,

(ii) the exposed-conductive-parts of the electrical equipment, and

(iii) the protective contacts of the socket-outlets.

608-03-03 Where the protective conductor specified in Regulation 608-03-02 is not incorporated in a cable and is not enclosed in conduit or trunking, it shall have a minimum cross-sectional area of 4 mm^2 and shall be insulated.

608-03-04 Except where the caravan or motor caravan is made substantially of insulating material, and metal parts are unlikely to become live in the event of a fault, extraneous-conductive-parts shall be bonded to the circuit protective conductor with a conductor of minimum cross-sectional area of 4 mm^2 and in more than one place if the construction of the caravan does not ensure continuity between extraneous-conductive-parts.

Metal sheets forming part of the structure of the caravan or motor caravan are not considered to be extraneous-conductive-parts.

Protection against overcurrent

608-04 Final circuits

608-04-01 Each final circuit shall be protected by an overcurrent protective device which disconnects all live conductors of that circuit.

Selection and erection of equipment

608-05   General

608-05-01 Where there is more than one electrically independent installation, each independent system shall be supplied by a separate connecting device and shall be segregated in accordance with Regulation 608-12.

608-06   Wiring systems

608-06-01 The following wiring systems shall be used with insulated conductors to the relevant British Standards:

(i)   flexible single-core insulated conductors in non-metallic conduits.

(ii) stranded insulated conductors in non-metallic conduits

(iii) sheathed flexible cables.

608-06-02 The cross-sectional area of every conductor shall not be less than 1.5 mm^2.

608-06-03 The limit of 6 mm^2 in Regulation 543-03-02 does not apply and all protective conductors regardless of cross-sectional area shall be insulated.

608-06-04 Cables of low voltage systems shall be run separately from the cables of extra-low voltage systems, in such a way, so far as is reasonably practicable, that there is no risk of physical contact between the two wiring systems.

608-06-05 All cables, unless enclosed in rigid conduit, and all flexible conduit shall be supported at intervals not exceeding 0.4 m for vertical runs and 0.25 m for horizontal runs.

608-06-06 No electrical equipment shall be installed in any compartment intended for the storage of fuel.

Switchgear and controlgear

608-07   Inlets

608-07-01 The electrical inlet to the caravan shall be an appliance inlet complying with BS 4343 of the two pole and earthing contact type with key position 6h.

608-07-02 The inlet shall be installed:

(i)   not more than 1.8 m above ground level,

(ii) in a readily accessible position,

(iii) in an enclosure with a suitable cover on the outside of the caravan.

608-07-03 A notice of such durable material as to be likely to remain easily legible throughout the life of the installation shall be fixed on or near the electrical inlet recess where it can be easily read and shall bear, in indelible and easily legible characters, the following information:

(i)   the nominal voltage and frequency for which the caravan installation concerned has been designed

(ii) the rated current of the caravan installation.

608-07-04 Every installation shall be provided with a main isolating switch which shall disconnect all live conductors and which shall be suitably placed for ready operation within the caravan. In an installation consisting of only one final circuit, the isolating switch may be the overcurrent protection device required in Regulation 608-04.

608-07-05 A notice of durable material shall be permanently fixed near the main isolating switch inside the caravan, bearing in indelible and easily legible characters the text shown below:

INSTRUCTIONS FOR ELECTRICITY SUPPLY

TO CONNECT

1. Before connecting the caravan installation to the mains supply, check that:

   (a) the supply available at the caravan pitch supply point is suitable for the caravan electrical installation and appliances, and

   (b) the caravan main switch is in the OFF position.

2. Open the cover to the appliance inlet provided at the caravan supply point and insert the connector of the supply flexible cable.

3. Raise the cover of the electricity outlet provided on the pitch supply point and insert the plug of the supply cable.

THE CARAVAN SUPPLY FLEXIBLE CABLE MUST BE FULLY UNCOILED TO AVOID DAMAGE BY OVERHEATING.

4. Switch on at the caravan main switch.

5. Check the operation of residual current devices, if any, fitted in the caravan by depressing the test button.

IN CASE OF DOUBT OR, IF AFTER CARRYING OUT THE ABOVE PROCEDURE THE SUPPLY DOES NOT BECOME AVAILABLE, OR IF THE SUPPLY FAILS, CONSULT THE CARAVAN PARK OPERATOR OR HIS AGENT OR A QUALIFIED ELECTRICIAN.

TO DISCONNECT

6. Switch off at the caravan main isolating switch, switch off at the pitch supply point and unplug both ends of the cable.

PERIODIC INSPECTION

Preferably not less than once every three years and more frequently if the vehicle is used more than normal average mileage for such vehicles, the caravan electrical installation and supply cable should be inspected and tested and a report on their condition obtained as prescribed in the Regulations for Electrical Installations published by the Institution of Electrical Engineers.

608-08   Accessories

608-08-01 Each accessory shall be of a type without accessible conductive parts.

608-08-02 Low voltage socket-outlets, other than those supplied by an individual winding of an isolating transformer, shall incorporate a protective contact.

608-08-03 All socket-outlets shall have their nominal voltage clearly and indelibly marked. Low voltage socket-outlets shall not be compatible with sockets supplied at extra-low voltage.

608-08-04 Where an accessory is located in a position in which it is exposed to the effects of moisture (AD5) it shall be constructed or enclosed so as to provide a degree of protection not less than IP 55.

608-08-05 Every appliance connected to the supply by a means other than a plug and socket shall be controlled by a switch connected to break all live conductors and incorporated in or adjacent to the appliance.

608-08-06 Each luminaire in a caravan shall preferably be fixed directly to the structure or lining of the caravan. Where a pendant luminaire is installed in a caravan, provision shall be made for securing the luminaire to prevent damage when the caravan is moved. A filament lamp luminaire shall be so designed or mounted as to allow a free circulation of air between it and the body of the caravan.

608-08-07 A luminaire intended for dual voltage operation shall:

(i) be fitted with separate lampholders for each voltage

(ii) have an indication of the lamp wattage and voltage clearly and permanently displayed near each lampholder

(iii) be so designed and constructed that no damage will be caused if both lamps are lit at the same time

(iv) be so designed and constructed that separation between conductors of different voltages is achieved

(v) be so designed that lamps cannot be inserted in lampholders intended for lamps of other voltages.

608-08-08 Every caravan having an electrical installation shall be equipped with a means of connection to the caravan pitch socket-outlet, comprising the following:

(i) a plug complying with BS 4343 and having key position 6h and being of the two pole and earthing contact type,

(ii) a flexible cord or cable not longer than 25 m, complying with BS 6007 or BS 6500, incorporating a protective conductor, and of a cross-sectional area in accordance with Table 608A and,

(iii) a connector complying with BS 4343 and being of the two pole and earthing contact type with key position 6h, and compatible with the appliance inlet installed under Regulation 608-07-01.

TABLE 608A

Cross-sectional areas of flexible cords and cables for caravan connectors

Rated current (A)	Cross-sectional area (mm^2)
16	2.5
25	4
32	6
63	16
100	35

SECTION 608

DIVISION TWO

ELECTRICAL INSTALLATIONS IN CARAVAN PARKS

608-09 Scope

608-09-01 The particular requirements of this Section apply to that portion of the electrical installation in caravan parks which provides facilities for the supply of electricity to, and connection of, leisure accommodation vehicles at nominal voltages not exceeding 250 V r.m.s., a.c. single phase.

PROTECTION AGAINST SHOCK

608-10 Protection against direct contact

608-10-01 The following methods of protection shall not be used:

(i) protection by obstacles (Regulations 412-04 and 471-06)

(ii) protection by placing out of reach (Regulations 412-05 and 471-07).

608-11 Protection against indirect contact

608-11-01 The following methods of protection shall not be used:

(i) non-conducting location (Regulations 413-04 and 471-10)

(ii) Earth free local equipotential bonding (Regulations 413-05 and 471-11)

(iii) electrical separation (Regulations 413-06 and 471-12).

Selection and erection of equipment

608-12 Wiring systems

608-12-01 So far as is practicable, caravan pitch supply equipment shall be connected by underground cable.

608-12-02 Underground cables, unless provided with additional mechanical protection, shall be installed outside any caravan pitch or other area where tent pegs or ground anchors may be driven.

608-12-03 All overhead conductors shall be:

(i) of a suitable construction and insulated in accordance with Regulation 412-02 and

(ii) located 2 m outside the vertical surface extending from the horizontal boundary of any caravan pitch

(iii) at a height of not less than 6 m in vehicle movement areas and 3.5 m in all other areas.

608-13 Switchgear and controlgear

608-13-01 Caravan pitch supply equipment shall be located adjacent to the pitch and not more than 20 m from any point on the pitch which it is intended to serve.

608-13-02 Each socket-outlet and its enclosure forming part of the caravan pitch supply equipment shall:

(i) be of the two pole and earthing contact type

(ii) comply with BS 4343, with key position 6h, and be of IPX4 to BS 5490

(iii) be placed at a height of between 0.80 m and 1.50 m from the ground to the lowest part of the socket-outlet

(iv) have a current rating of 16 A

(v) have not less than one 16 A socket-outlet provided for each pitch.

608-13-03 Where demands greater than 16 A are intended to be supplied, additional socket-outlets with higher ratings shall be installed.

608-13-04 Each socket-outlet shall be individually protected by an overcurrent device.

608-13-05 Socket-outlets shall be protected individually, or in groups of not more than six, by a residual current device complying with BS 4293 and having the characteristics specified in Regulation 412-06 and must not be bonded to the PME terminal.

For a PME supply the protective conductor of each socket-outlet circuit shall be connected to an earth electrode and shall comply with Regulations 413-02-18 to 413-02-20.

608-13-06 Grouped socket-outlets shall be on the same phase.

SECTION 609

(Reserved for Marinas)

SECTION 610

(Reserved for future use)

SECTION 611

HIGHWAY POWER SUPPLIES AND STREET FURNITURE

611-01   Scope

611-01-01   The requirements of this Section shall apply to installations comprising highway distribution circuits, street furniture and other street located equipment. They shall not apply to supplier's works as defined by the Electricity Supply Regulations 1988 (as amended) in accordance with Regulation 110-02.

611-01-02   Any measure prescribed in this Section of the Regulations shall also apply to similar equipment located in other areas used by the public but not designated as a highway or part of a building.

Protection for safety

611-02   Protection against electric shock

611-02-01   Where a measure for protection against direct contact in accordance with Regulation 412-01 is used then:

(i)   protection by obstacles shall not be used, and

(ii)   where protection is provided by placing out of reach, it shall only apply to low voltage overhead lines constructed to the standard required by the Electricity Supply Regulations 1988 (as amended)

(iii) except when the maintenance of equipment is to be restricted to skilled persons specially trained, where items of street furniture or street located equipment are within 1.5 m of a low voltage overhead line, protection against direct contact with the overhead line shall be provided by means other than placing out of reach.

611-02-02   A door in street furniture or street located equipment used for access to electrical equipment shall not be used, to meet Regulation 412-03-01, as a barrier or enclosure. To satisfy the purposes of protection against direct contact the requirements of Regulation 412-03-04 shall be applied. An intermediate barrier shall be provided to prevent contact with live parts, such barrier affording a degree of protection of at least 1P2X (see BS 5490) and removable only by the use of a tool.

611-02-03   The measure for protection against indirect contact in accordance with Regulation 413-01 shall not be taken from the following:

(i)   non-conducting location

(ii)   earth-free equipotential bonding

(iii) electrical separation

611-02-04   A maximum disconnection time of 5 s shall apply to all circuits feeding fixed equipment used in highway power supplies for compliance with the requirements of Regulation 413-02-04.

611-02-05   Where protection against indirect contact is provided by using earthed equipotential bonding and automatic disconnection in accordance with Regulation 413-01-01(i) metallic structures not connected to or part of the street furniture or street located equipment shall not be connected to the main earthing terminal as extraneous conductive parts under Regulation 413-02-02.

611-03   Devices for isolation and switching

611-03-01   For an item of street furniture or street located equipment including an item used as a highway distribution board for supplies to other equipment Regulation 460-01-02 shall only apply where more than one highway distribution circuit is connected or the maximum load current exceeds either 16 A or the switching rating of the device.

611-03-02   Where the supplier's cut-out is used as the means of isolation of a highway power supply the approval of the supplier shall be obtained. The electrical maintenance of installations where the exemption in Regulation 611-03-01 applies shall be restricted to skilled persons and/or instructed persons.

Selection and erection of equipment

611-04 Identification of cables

611-04-01 On completion of an installation including highway distribution circuits and highway power supplies, detailed records in accordance with Regulation 514-09 shall be provided with the Completion and Inspection Certificate required by Regulation 741-01-01.

611-04-02 Except where the method of cable installation does not permit marking the installation of underground cable shall comply with Regulation 522-06-03.

611-04-03 Ducting, marker tape or cable tiles used with highway power supply cable shall be suitably colour coded or marked for the purpose of identification and shall be distinct from other services.

611-04-04 The requirement of Regulation 514-12 need not be applied where the highway power supply installation is subject to a programmed Inspection and Testing procedure.

611-05 External influences

611-05-01 Where conductors are installed in conduit or duct within the ground the system shall comply with the requirement for mechanical protection in Section 522.

611-06 Temporary supplies

611-06-01 Temporary supplies taken from street furniture shall not reduce the safety of the permanent installation and shall generally be in accordance with Section 604.

611-06-02 On every temporary supply unit there shall be a durable label externally mounted stating the maximum sustained current to be supplied from that unit.

PART 7

INSPECTION AND TESTING

CONTENTS

CHAPTER 71   **INITIAL VERIFICATION**

711   GENERAL

712   INSPECTION

713   TESTING

CHAPTER 72   **ALTERATIONS AND ADDITIONS TO INSTALLATIONS**

721   GENERAL

CHAPTER 73   **PERIODIC INSPECTION AND TESTING**

731   GENERAL

732   INSPECTION AND TESTING

CHAPTER 74   **CERTIFICATION AND REPORTING**

741   GENERAL

742   INITIAL VERIFICATION

743   ALTERATIONS AND ADDITIONS

744   PERIODIC INSPECTION AND TESTING

PART 7

INSPECTION AND TESTING

CONTENTS

CHAPTER 71	**INITIAL VERIFICATION**	
711	GENERAL	
712	INSPECTION	
713	TESTING	
713-01	General	
713-02	Continuity of protective conductors	
713-03	Continuity of ring final circuit conductors	
713-04	Insulation resistance	
713-05	Site-applied insulation	
713-06	Protection by separation of circuits	
713-07	Protection against direct contact, by a barrier or enclosure provided during erection	
713-08	Insulation of non-conducting floors and walls	
713-09	Polarity	
713-10	Earth fault loop impedance	
713-11	Earth electrode resistance	
713-12	Operation of residual current operated devices	
CHAPTER 72	**ALTERATIONS AND ADDITIONS TO INSTALLATIONS**	
721	GENERAL	
CHAPTER 73	**PERIODIC INSPECTION AND TESTING**	
731	GENERAL	
732	INSPECTION AND TESTING	

CHAPTER 74   CERTIFICATION AND REPORTING

741   GENERAL

742   INITIAL VERIFICATION

743   ALTERATIONS AND ADDITIONS

744   PERIODIC INSPECTION AND TESTING

CHAPTER 71

INITIAL VERIFICATION

711 GENERAL

711-01-01 Every installation shall, during erection and/or on completion before being put into service be inspected and tested to verify, so far as is reasonably practicable, that the requirements of the Regulations have been met.

The method of test shall be such that no danger to persons, livestock or property or damage to equipment can occur even if the circuit tested is defective.

711-01-02 The result of the assessment of general characteristics required by Sections 311, 312 and 313, together with the information required by Regulation 514-09-01 shall be made available to the person or persons carrying out the inspection and testing.

712 INSPECTION

712-01-01 Detailed inspection shall precede testing and shall normally be done with that part of the installation under inspection disconnected from the supply.

712-01-02 The detailed inspection shall be made to verify that the installed electrical equipment is:

(i) in compliance with Section 511 (this may be ascertained by mark or by certification furnished by the installer or by the manufacturer), and

(ii) correctly selected and erected in accordance with the Regulations, and

(iii) not visibly damaged or defective so as to impair safety.

712-01-03 The detailed inspection shall include at least the checking of the following items, where relevant to the installation and, where necessary, during erection:

(i) connection of conductors

(ii) identification of conductors

(iii) routing of cables in safe zones or mechanical protection, in compliance with Section 522

(iv) selection of conductors for current-carrying capacity and voltage drop, in accordance with the design

(v) connection of single pole devices for protection or switching in phase conductors only

(vi) correct connection of socket-outlets and lampholders

(vii) presence of fire barriers and protection against thermal effects

(viii) methods of protection against direct contact (including measurement of distances where appropriate), i.e.:

  a - protection by insulation of live parts

  b - protection by barrier or enclosure

  c - protection by obstacles

  d - protection by placing out of reach

(ix) methods of protection against indirect contact:

    a - presence of protective conductors

    b - presence of earthing conductors

    c - presence of main equipotential bonding conductors

    d - presence of supplementary equipotential bonding conductors

    e - earthing arrangements for combined protective and functional purposes

    f - use of Class II equipment or equivalent insulation

    g - non-conducting location (including measurement of distances, where appropriate)

    h - earth-free local equipotential bonding

    i - electrical separation

(x) prevention of mutual detrimental influence

(xi) presence of appropriate devices for isolation and switching

(xii) presence of undervoltage protective devices

(xiii) choice and setting of protective and monitoring devices (for protection against indirect contact and/or protection against overcurrent)

(xiv) labelling of circuits, fuses, switches and terminals

(xv) selection of equipment and protective measures appropriate to external influences

(xvi) adequacy of access to switchgear and equipment

(xvii) presence of danger notices and other warning notices

(xviii) presence of diagrams, instructions and similar information

(xix) erection methods.

713 TESTING

713-01 General

713-01-01 The tests of Regulations 713-02 to 713-09 where relevant shall be carried out in that sequence.

In the event of any test indicating failure to comply, that test and those preceding, the results of which may have been influenced by the fault indicated, shall be repeated after the fault has been rectified.

Reference methods of test are described in Guidance Notes on the Wiring Regulations published by the Institution of Electrical Engineers, but the use of other methods giving no less effective results is not precluded.

713-02 Continuity of protective conductors

713-02-01 Every protective conductor shall be tested to verify that it is electrically sound and correctly connected.

713-03 Continuity of ring final circuit conductors

713-03-01 A test shall be made to verify the continuity of each conductor including the protective conductor, of every ring final circuit.

713-04 Insulation resistance

713-04-01 The insulation resistance between live conductors shall be measured, before the installation is connected to the supply.

713-04-02 Particular attention shall be given to the presence of electronic devices connected in the installation and, where necessary, such devices shall be isolated so that they are not damaged by the test voltage and thereafter tested in accordance with Regulation 713-04-06.

713-04-03 The insulation resistance shall also be measured between each live conductor and earth, the PEN conductor in TN-C systems being considered as part of the earth. Where appropriate during this measurement, phase and neutral conductors may be connected together.

713-04-04 The insulation resistance measured with the d.c. test voltages indicated in Table 71A shall be considered satisfactory if the main switchboard, and each distribution circuit tested separately with all its final circuits connected but with current-using equipment disconnected, has an insulation resistance not less than the appropriate value given in Table 71A.

713-04-05 The testing equipment shall be capable of supplying the test voltage indicated in Table 71A when loaded with 1 mA.

TABLE 71A
Minimum values of insulation resistance

Circuit nominal voltage (volts)	Test voltage d.c. (volts)	Minimum insulation resistance (megohms)
Extra-low voltage circuits when the circuit is supplied from a safety isolating transformer (Regulation 411-02-02 item (i)) and also fulfils the requirements of Regulation 411-02-04	250	0.25
Up to and including 500 V with the exception of the above cases	500	0.5
Above 500 V up to 1000 V	1000	1.0
Between SELV circuits and associated LV circuits	500	5.0
with additional withstand test where required e.g. U_0 240 V		3750 r.m.s. a.c. for 60 s

713-04-06 Where equipment, such as electronic devices, is disconnected for the tests prescribed in Regulations 713-04-01 to 05, and the equipment has exposed-conductive-parts required by the Regulations to be connected to protective conductors, the insulation resistance between the exposed-conductive-parts and all live parts of the disconnected equipment shall be measured separately and shall comply with the requirements of the appropriate British Standard for the equipment. If there is no appropriate British Standard the insulation resistance shall be not less than 0.5 megohm.

713-05 Site applied insulation

713-05-01 Where insulation applied on site in accordance with Regulation 412-02 is intended to provide protection against direct contact, it shall be verified that the insulation is capable of withstanding, without breakdown or flashover, an applied voltage test equivalent to that specified in the British Standard for similar type-tested equipment.

713-05-02 Where protection against indirect contact is provided by supplementary insulation applied to equipment during erection in accordance with Regulation 413-03, it shall be verified by test:

(i) that the insulating enclosure affords a degree of protection not less than IP2X (BS 5490), and

(ii) that the insulating enclosure is capable of withstanding, without breakdown or flashover, an applied voltage test equivalent to that specified in the British Standard for similar type tested equipment.

713-06 Protection by separation of circuits

713-06-01 Where protection against electric shock is provided by SELV, compliance with the requirements of Regulations 411-02 and 471-02 shall be verified by inspection and test.

713-06-02 Where protection against electric shock is provided by electrical separation, compliance with the requirements of Regulation 413-06 shall be verified by inspection and test.

713-07 Protection against direct contact, by a barrier or enclosure provided during erection

713-07-01 Where protection against direct contact is intended to be afforded by a barrier or enclosure provided during erection in accordance with Regulation 412-03, it shall be verified by test that each enclosure or barrier affords a degree of protection not less than IP2X or IP4X as appropriate, where that Regulation so requires.

713-08 Insulation of non-conducting floors and walls

713-08-01 Where protection against indirect contact is to be provided by a non-conducting location intended to comply with Regulations 413-04 and 471-10, the resistance of the floors and walls of the location to the main protective conductor of the installation shall be measured at not less than three points on each relevant surface, one of which shall be not less than 1 m and not more than 1.2 m from any extraneous-conductive-part in the location. The other two measurements shall be made at greater distances.

713-08-02 Any insulation or insulating arrangement of extraneous-conductive-parts intended to satisfy Regulation 413-04-07 (iii):

(i) when tested at 500 V d.c. shall not be less than 0.5 megohm, and

(ii) shall be able to withstand a test voltage of at least 2 kV r.m.s., a.c. and

(iii) shall not pass a leakage current exceeding 1 mA in normal conditions of use.

713-09 Polarity

713-09-01 A test of polarity shall be made and it shall be verified that:

(i) every fuse and single-pole control and protective device is connected in the phase conductor only

(ii) centre-contact bayonet and Edison screw lampholders to BS 6776 in circuits having an earthed neutral conductor have the outer or screwed contacts connected to the neutral conductor

(iii) wiring has been correctly connected to socket-outlets and similar accessories.

713-10 Earth fault loop impedance

713-10-01 Where protective measures are used which require a knowledge of earth fault loop impedance, the relevant impedances shall be measured, or determined by an alternative method.

Where the alternative method described in Regulation 413-02-12 (see Table 41C) is used, the impedance of the protective conductor of the circuit concerned shall also be measured.

713-11 Earth electrode resistance

713-11-01 Where protective measures are used which require a knowledge of the earth electrode resistance, this shall be measured.

713-12 Operation of residual current operated devices

713-12-01 Where protection against indirect contact is to be provided by a residual current device its effectiveness shall be verified by a test simulating an appropriate fault condition and independent of any test facility incorporated in the device.

CHAPTER 72

ALTERATIONS AND ADDITIONS TO AN INSTALLATION

721 GENERAL

721-01-01 The relevant requirements of Chapter 71 shall apply to alterations and additions.

721-01-02 It shall be verified that every alteration or addition complies with the Regulations and does not impair the safety of an existing installation.

CHAPTER 73

PERIODIC INSPECTION AND TESTING

731 GENERAL

731-01-01 Periodic Inspection and Testing of every installation where required, shall be carried out in accordance with the requirements of this Chapter.

731-01-02 Inspection comprising careful scrutiny of the installation shall be carried out without dismantling or with partial dismantling as required, supplemented by testing to verify compliance with Sections 731 and 732 and as far as possible to provide for:

(i) the safety of persons and livestock against the effects of electric shock and burns, in accordance with Regulation 120-01, and

(ii) protection against damage to property by fire and heat arising from an installation defect, and

(iii) the identification that the installation is not damaged or deteriorated so as to impair safety, and

(iv) the identification of installation defects or non- compliance with the requirements of the Regulations, which may give rise to danger.

732 INSPECTION AND TESTING

732-01-01 The frequency of Periodic Inspection and Testing of an installation shall be determined by the type of installation, its use and operation, the frequency of maintenance and the external influences to which it is subjected.

732-01-02 The Inspection and Testing shall not cause danger to persons or livestock and shall not cause damage to property and equipment even if the circuit is defective.

732-01-03 The results of Periodic Inspection and Testing shall be recorded on a report, signed by the person carrying out the Inspection or a person authorised to act on his behalf and forwarded to the originator of the request for the Inspection in accordance with Section 744 of the Regulations.

CHAPTER 74

CERTIFICATION AND REPORTING

741 GENERAL

741-01-01 Following the initial verification required by Chapter 71 or Chapter 72, a Completion and Inspection Certificate in the form set out in Appendix 6 shall be given to the person ordering the work.

741-01-02 Following the Periodic Inspection and Testing described in Chapter 73 a Report on the findings of the periodic inspection and testing shall be given by the person carrying out the inspection, or by a person authorised to act on his behalf, to the person ordering the work.

742 INITIAL VERIFICATION

742-02-01 The Inspection shall comply with the requirements of Chapter 71 and any defects or omissions revealed by the Inspection shall be made good before a Completion and Inspection Certificate is issued.

742-02-02 The Completion and Inspection Certificate shall be signed by a competent person or persons stating that the installation has been designed, constructed and inspected and tested in accordance with the Regulations.

743 ALTERATIONS AND ADDITIONS

743-01-01 The requirements of Section 742 for the issue of a Completion and Inspection Certificate shall apply to all the work of the alterations or additions. Any defects or omissions revealed in that work shall be made good before a Completion Report is issued.

The Contractor or other person responsible for the new work, or a person authorised to act on his behalf, shall report in writing, to the person ordering the work any defects found in related parts of the existing installation.

744 PERIODIC INSPECTION AND TESTING

744-01-01 The results and extent of a Periodic Inspection and Testing, in accordance with Chapter 73, of an installation, or any part thereof, shall be recorded on a report and given by the person carrying out the inspection or a person authorised to act on his behalf, to the person ordering the inspection.

744-01-02 Dangerous conditions arising from non-compliance with these Regulations together with any limitations of the Inspection and Testing in accordance with Section 732, shall be recorded.

Appendices

CONTENTS

Appendix		page
1	British Standards to which reference is made in the Regulations	152
2	Statutory Regulations and associated memoranda	158
3	Time/Current characteristics of overcurrent protective devices	160
4	Current-carrying capacity and voltage drop for cables and flexible cords	171
5	Classification of external influences	240
6	Forms of Completion and Inspection Certificate	242

APPENDIX 1

BRITISH STANDARDS TO WHICH REFERENCE IS MADE IN THE REGULATIONS

BS Number	Title	Referenced in Regulations
BS 31 : 1940(1980)	Specification. Steel conduit and fittings for electrical wiring.	521-04-01(i)
BS 67 : 1987	Specification for ceiling roses.	553-04-01(i)
BS 88 :	Cartridge fuses for voltages up to and including 1000 V a.c. and 1500 V d.c. Part 2 : Specification for fuses for use by authorized persons (mainly for industrial applications). Part 6 : 1988 Specification of supplementary requirements for fuses of compact dimensions for use in 240/415 V a.c. industrial and commercial electrical installations.	433-02-02 Table 41B1 Table 41C Table 41D Table 471A Table 604B1
BS 196 : 1973(1988)	Specification for protected-type non-reversible plugs, socket-outlets cable-couplers and appliance-couplers with earthing contacts for single phase a.c. circuits up to 250 volts.	Table 55A 553-01-05 553-02-01
BS 415 : 1979(1987)	Specification for safety requirements for mains-operated electronic and related apparatus for household and similar general use.	471-14-02
BS 476 :	Fire tests on building materials and structure.	521-05-01
	Part 4 : 1970 (1984) Non-combustibility test for materials.	526-03-02(iv)
	Part 5 : 1979 Method of test for ignitability.	526-03-02(v) 527-01-03 553-03-01
	Part 23 : 1987 Methods for determination of the contribution of components to the fire resistance of a structure.	527-02-02
BS 546 : 1950(1988)	Specification. Two-pole and earthing pin plugs, socket-outlets and socket-outlet adaptors.	Table 55A 553-01-05
BS 559 : 1986	Specification for electric signs and high-voltage luminous-discharge-tube installation.	110-01-01(ix) 554-02-01
BS 646 : 1958(1986)	Specification. Cartridge fuse links (rated up to 5 amperes) for a.c. and d.c. service.	Table 55A 553-01-05(i)
BS 731 :	Flexible steel conduit for cable protection and flexible steel tubing to enclose flexible drives. Part 1 : 1952 (1980) Flexible steel conduit and adaptors for the protection of electric cables.	521-04-01(ii)

BS Number	Title	Referenced in Regulations
BS 1361 : 1971(1986)	Specification for cartridge fuses for a.c. circuits in domestic and similar premises.	Table 41B1 Table 41C Table 41D Table 604B1 433-02-02
BS 1362 : 1973(1986)	Specification for general purpose fuse links for domestic and similar purposes (primarily for use in plugs).	Table 41B1 Table 41C Table 41D Table 55A Table 604B1 553-01-05(i)
BS 1363 : 1984	Specification for 13 A fused plugs and switched and unswitched socket-outlets.	Table 55A 553-01-04 553-01-05
BS 1710 : 1984(1989)	Specification for identification of pipelines and services.	514-02-01
BS 2632 : 1980(1986)	Specification for mains-operated electric fence controllers.	471-03-01 605-14-01
BS 2754 : 1976	Memorandum. Construction of electrical equipment for protection against electric shock.	Part 2
BS 2848 : 1973	Specification for flexible insulating sleeving for electrical purposes.	543-03-02
BS 3036 : 1958(1986)	Specification. Semi-enclosed electric fuses (rating up to 100 amperes and 240 volts to earth).	Table 41B1 Table 41D Table 604B1 433-02-03 APP 3 APP 4
BS 3456 :	Specification for safety of household and similar electrical appliances Section 3.9 : 1979 Stationary instantaneous water heaters.	601-08-01(iii) 602-08-03
BS 3535 :	Isolating transformers and safety transformers. Part 1 : 1990 General requirements. Part 2 : 1990 Specification for transformers for reduced system voltage.	411-02-02(i) 413-06-02(i) (a) 413-06-03(iv) 471-12-01 471-15-03(i) 553-01-05(ii) 601-09-01 602-07-02(iv)
BS 3676 :	Switches for household and similar fixed electrical installations. Part 1 : 1989 Specification for general requirements.	412-03-04(iii) 553-04-01 601-08-01(i) 602-07-02 (iv)
BS 3858 : 1965	Specification for binding and identification sleeve for use on electric cables and wires.	514-06-01
BS 3871 :	Specification for miniature and moulded case circuit-breakers. Part 1 : 1965 (1984) Miniature air-break circuit-breakers for a.c. circuits.	Table 41B2 Table 41C Table 471A Table 604B2

BS Number	Title	Referenced in Regulations
BS 3939 :	Guide for graphical symbols for electrical power, telecommunications and electronic diagrams.	514-09-01
BS 4099 :	Colours of indicator lights, push-buttons, annunciators and digital readouts.	514-01-01
BS 4293 : 1983	Specification for residual current-operated circuit-breakers.	412-06-02(ii)
BS 4343 : 1968	Specification for industrial plugs, socket-outlets and couplers for a.c. and d.c. supplies.	Table 55A 553-01-05 553-02-01 602-08-01 604-12-02 604-13-01 607-02-04 607-02-05 608-07-01 608-08-08(i) 608-08-08(iii) 608-13-02(ii)
BS 4363 : 1991	Specification for distribution assemblies for electricity supplies for construction and building sites.	604-09-01
BS 4444 : 1989	Guide to electrical earth monitoring and protective conductor proving.	543-03-05
BS 4491 :	Appliance couplers for household and similar general purposes.	553-02-01
BS 4533 : 1990	Section 102.57 Electrical supply track systems for luminaires.	521.06.01
BS 4568 : 1970	Part 1. Steel conduit, bends and couplers.	521-04-01(iii)
BS 4573 : 1979	Specification for 2-pin reversible plugs and shaver socket-outlets.	553-01-05(ii)
BS 4607	Part 2: 1985 Rigid p.v.c. conduits and conduit fittings. Part 3: 1985 Pliable corrugated, plain and reinforced conduits of self-extinguishing plastics material. Part 5: 1988 Specification for rigid conduits, fittings and components of insulating material.	521-04-01(iv)
BS 4678	Cable trunking.	521-05-01
BS 4727	Glossary of electrotechnical, power, telecommunications, electronics, lighting and colour terms.	Part 2

BS Number	Title	Referenced in Regulations
BS 4941	Specification for motor starters for voltages up to and including 1000 V a.c. and 1200 V d.c.	435-01-01
BS 5042 : 1987	Specification for bayonet lampholders.	412-03-04 471-05-02 Table 55B 553-03-03 553-04-01(iii) 601-11-01
BS 5266 :	Emergency lighting.	110-01-01(x) 528-01-04
BS 5345 :	Code of practice for selection, installation and maintenance of electrical apparatus for use in potentially explosive atmospheres (other than mining applications or explosive processing and manufacture).	110-01-01(xi)
BS 5467 : 1989	Specification for cables with thermosetting insulation for electricity supply for rated voltages of up to and including 600/1000 V and up to and including 1900/3300 V.	514-06-01(iv)
BS 5486 :	Low-voltage switchgear and controlgear assemblies. Part 1 : 1990 Requirements for type-tested and partially type-tested assemblies. Part 2 : 1988 Particular requirements for busbar trunking systems (busways). Part 4 : 1991 Particular requirements for assemblies for construction sites (ACS).	413-03-01(i)b 521-01-02 543-02-04(ii) 604-09-01
BS 5490 : 1977(1985)	Specification for classification of degrees of protection provided by enclosures.	411-02-09(i) 412-03-01 412-03-02 412-03-04(iii) 412-05-03 413-03-04 471-14-02(i) 602-05-01 603-03-01(ii) 608-13-02(ii) 611-02-02 713-05-02(i)
BS 5518 : 1977	Specification for electronic variable control switches (dimmer switches) for tungsten filament lighting.	553-04-01
BS 5593 : 1978(1990)	Specification for impregnated paper-insulated cables with aluminium sheath/neutral conductor and three shaped solid aluminium phase conductors (CONSAC), 600/1000 V, for electricity supply.	546-02-04(ii)
BS 5655 :	Lifts and service lifts.	110-02-01(x) 528-02-06

BS Number	Title	Referenced in Regulations
BS 5839 :	Fire detection and alarm systems for buildings.	110-01-01(xii) 528-01-04
BS 6004 : 1990	Specification for p.v.c.-insulated cables (non-armoured) for electric power and lighting.	543-03-02
BS 6007 : 1983	Specification for rubber-insulated cables for electric power and lighting.	608-08-08(ii)
BS 6053 : 1981(1990)	Specification for outside diameters of conduits for electrical installations and threads for conduit and fittings.	521-04-01(i) 521-04-01(ii) 521-04-01(iii)
BS 6099 :	Conduits for electrical installations. Part 1 : 1981 (1986) Specification of general requirements. Section 2.2 : 1982 (1988) Specification for rigid plain conduits of insulating material.	521-04-01(i) 521-04-01(ii) 521-04-01(iii) 521-04-01(iv)
BS 6141 : 1991	Specification for insulated cables and flexible cords for use in high temperature zones.	603-07-01
BS 6207 : 1987	Specification for mineral-insulated copper-sheathed cables with copper conductors.	546-02-07
BS 6231 : 1981	Specification for p.v.c.-insulated cable for switchgear and controlgear wiring.	APP4
BS 6346 : 1989	Specification for p.v.c.-insulated cables for electricity supply.	514-06-01(ii)
BS 6351 :	Electric surface heating. Part 1 : 1983 Specification for electric surface heating devices. Part 2 : 1983 Guide to the design of electric surface heating systems. Part 3 : 1983 Code of practice for the installation, testing and maintenance of electric surface heating systems.	110-01-01(xiv) 554-07-01
BS 6369 : 1983	Specification for battery-operated electric fence controllers suitable for connection to the supply mains.	605-14-01
BS 6387 : 1983	Specification for performance requirements for cables required to maintain circuit integrity under fire conditions.	528-01-06
BS 6458 :	Fire hazard testing for electrotechnical products. Section 2.1 : 1984 Glow-wire test.	526-03-02(iii)
BS 6480 : 1988	Specification for impregnated paper-insulated lead or lead alloy sheathed electric cables of rated voltages up to and including 33000 V.	514-06-01(iii)
BS 6500 : 1990	Specification for insulated flexible cords and cables.	608-08-08(ii)

BS Number	Title	Referenced in Regulations
BS 6651 : 1985	Code of practice for protection of structures against lightning.	110-02-01(ix) 541-01-03
BS 6701 :	Code of practice for installation of apparatus intended for connection to certain telecommunications systems. Part 1 : 1986 General recommendations.	110-01-01(xiii) Table 51A 528-01-03 528-01-04
BS 6702 : 1986	Specification for lampholders for Tubular fluorescent lamps and starterholders.	Table 55B
BS 6724 : 1986	Specification for armoured cables for electricity supply having thermo-setting insulation with low emission of smoke and corrosive gases when affected by fire.	APP4
BS 6765 :	Leisure accommodation vehicles: caravans. Part 3 : 1989 specification for 12 V direct current extra-low voltage electrical installations.	608-01-01
BS 6776 :	Specification for edison screw lampholders.	412-03-04 Table 55B 553-04-01(iii) 713-09-01(ii)
BS 6883 : 1991	Specification for elastomer insulated cables for fixed wiring in ships.	APP4
BS 6972 : 1988	Specification for general requirements for luminaire supporting couplers for domestic, light industrial and commercial use.	553-04-01(ii)
BS 6991 : 1990	6/10 amp two pole weather-resistant couplers for household, commercial and light industrial equipment.	553-02-01
BS 7001 : 1988	Specification for interchangeability and safety of a standardized luminaire supporting coupler.	553-04-01(ii)
BS 7002 : 1989	Specification for safety of information technology equipment including electrical business equipment.	607-02-02
BS 7211 : 1989	Specification for thermosetting insulated cables (non-armoured) for electric power and lighting with low emission of smoke and corrosive gases when affected by fire.	543-03-02
BS 7454 : 1991	Calculation of thermally permissible short-circuit currents taking into account non-adiabatic heating effects.	434-03-03 543-01-03
BS EN 60 947-3	Low-voltage switchgear and controlgear. Part 3 : Switches, disconnectors, switch-disconnectors & fuse-combination units.	537-02-01

APPENDIX 2

STATUTORY REGULATIONS AND ASSOCIATED MEMORANDA

1. In Great Britain the following classes of electrical installations are required to comply with the statutory regulations indicated below. The regulations listed represent the principal legal requirements. Information concerning these regulations may be obtained from the appropriate authority also indicated below.

Provisions relating to electrical installations are also to be found in other legislation relating to particular activities.

(i)	Installations generally, subject to certain exemptions.	Electricity Supply Regulations 1988 (as amended) (SI 1988 No 1057 : ISBN 011 087057 3) (SI 1990 No 390 : ISBN 011 003390 6)	Secretary of State for Energy and Secretary of State for Scotland.
(ii)	Building generally (for Scotland only), subject to certain exemptions.	Building Standards (Scotland) Regulations 1990	Secretary of State for Scotland.
(iii)	Non-domestic installations. Places of work activity.	Electricity at Work Regulations 1989. (SI 1989 No 635 : ISBN 011 05635 X)	Health and Safety Commission.
(iv)	Cinematograph installations.	Cinematograph Regulations 1955, made under the Cinematograph Act, 1909, and/or Cinematograph Act, 1952.	H.M. Fire Service Inspectorate. Home Office, and Secretary of State for Scotland.
(v)	Agricultural and horticultural installations.	Agricultural (Stationary Machinery) Regulations 1959 as amended. (SI 1959 No 1216 1776 No 1247) (SI 1981 No 1414)	Health and Safety Commission.

2. Failure to comply in a consumer's installation in Great Britain with the requirements of Chapter 13 of the IEE Regulations For Electrical Installations places the supplier in the position of not being compelled to commence or, in certain circumstances, to continue to give, a supply of energy to that installation.

Under Regulation 29 of the Electricity Supply Regulations 1988 (as amended), any difference which may arise between a consumer and the supplier having reference to the consumer's installation shall be determined, in England and Wales, by a person nominated by the Secretary of State on the application of the consumer or his authorised agent or the supplier.

3. Where it is intended to use Protective Multiple Earthing the supplier and the consumer must comply with the Electricity Supply Regulations 1988 (as amended).

4. For further guidance on the application of some other of the above mentioned statutory regulations, reference may be made to the following publication:

   (i) Memorandum of Guidance on the Electricity at Work Regulations 1989. (HS(R)25) ISBN 011 883963 2.

5. For installations in potentially explosive atmospheres reference should be made to:

(i) the Electricity at Work Regulations 1989

(ii) the Highly Flammable Liquids and Liquified Petroleum Gases Regulations 1972

(iii) the Petroleum (Consolidation) Act 1928

(iv) relevant British Standards.

Under the Petroleum (Consolidation) Act local authorities are empowered to grant licences in respect of premises where petroleum spirit is stored and as the authorities may attach such conditions as they think fit, the requirements may vary from one local authority to another. Guidance may be obtained from the Health and Safety Executive (Guidance Note HS(G)41. Petrol filling Stations : Instructions and Operation).

6. For installations in theatres and other places of public entertainment, and on caravan sites, the requirements of the licensing authority should be ascertained. Model Standards were issued by the Department of Environment in 1977 under the Caravan Sites and Control of Development Act 1960 as guidance for local authorities.

7. The Low Voltage Electrical Equipment (Safety) Regulations 1989 (SI 1989 No 728 : ISBN 011 096728 3), administered by the Department of Trade and Industry, contain requirements for the safety of equipment designed or suitable for general use. Information on the application of the regulations is given in the booklet 'Administrative Guidance on the Electrical Equipment (Safety) Regulations 1975 and the Electrical Equipment (Safety) (Amendment) Regulations 1976'.

		Conditions of licence under:	
(i)	Theatres and other places licensed for public entertainment, music, dancing, etc.	a) in England and Wales The Local Government (Miscellaneous provisions) Act 1982	a) Home Office
		b) in Scotland the Civic Government (Scotland) Act 1982	b) Secretary of State for Scotland.
(ii)	High voltage luminous tube signs	As (a) and (b) above	As (a) and (b) above

8. The Plugs and Sockets etc. (Safety) Regulations 1987 (SI 1987 No 603 : ISBN 011 076603 2) of the Consumer Safety Act 1978, administered by the Department of Trade and Industry, containing requirements for the safety of plugs, sockets, adaptors and fuse links etc. designed for domestic use at a voltage of not less than 50 volts.

9. Where a pictographic safety sign is used for a caution of risk of electric shock, the Safety Signs Regulations (SI 1980 No 1471), administered by the Health and Safety Executive, are applicable.

APPENDIX 3

TIME/CURRENT CHARACTERISTICS OF OVERCURRENT PROTECTIVE DEVICES

This appendix gives the time/current characteristics of the following overcurrent devices:

Figure 1   HBC fuses to BS 1361

Figures 2A & 2B   Semi-enclosed fuses to BS 3036

Figures 3A & 3B   HBC fuses to BS 88 Part. 2 and Part. 6

Miniature circuit-breakers (m.c.bs.) to BS 3871

Figure 4   Type 1

Figure 5   Type 2

Figure 6   Type 3

Figure 7   Type B

Figure 8   Type C

In all of these cases time/current characteristics are based on the slowest operating times for compliance with the Standard and have been used as the basis for determining the limiting values of earth fault loop impedance prescribed in Section 413 of Chapter 41.

Time/Current characteristics for fuses to BS 1361

FUSE RATING	CURRENT FOR TIME			
	0.1 sec	0.2 sec	0.4 sec	5 secs
5 A	30 A	25 A	22 A	14 A
15 A	97 A	80 A	70 A	46 A
20 A	180 A	155 A	135 A	82 A
30 A	280 A	240 A	200 A	125 A
45 A	550 A	470 A	400 A	240 A
60 A	880 A	720 A	600 A	330 A
80 A	1100 A	950 A	800 A	460 A
100 A	1800 A	1400 A	1200 A	630 A

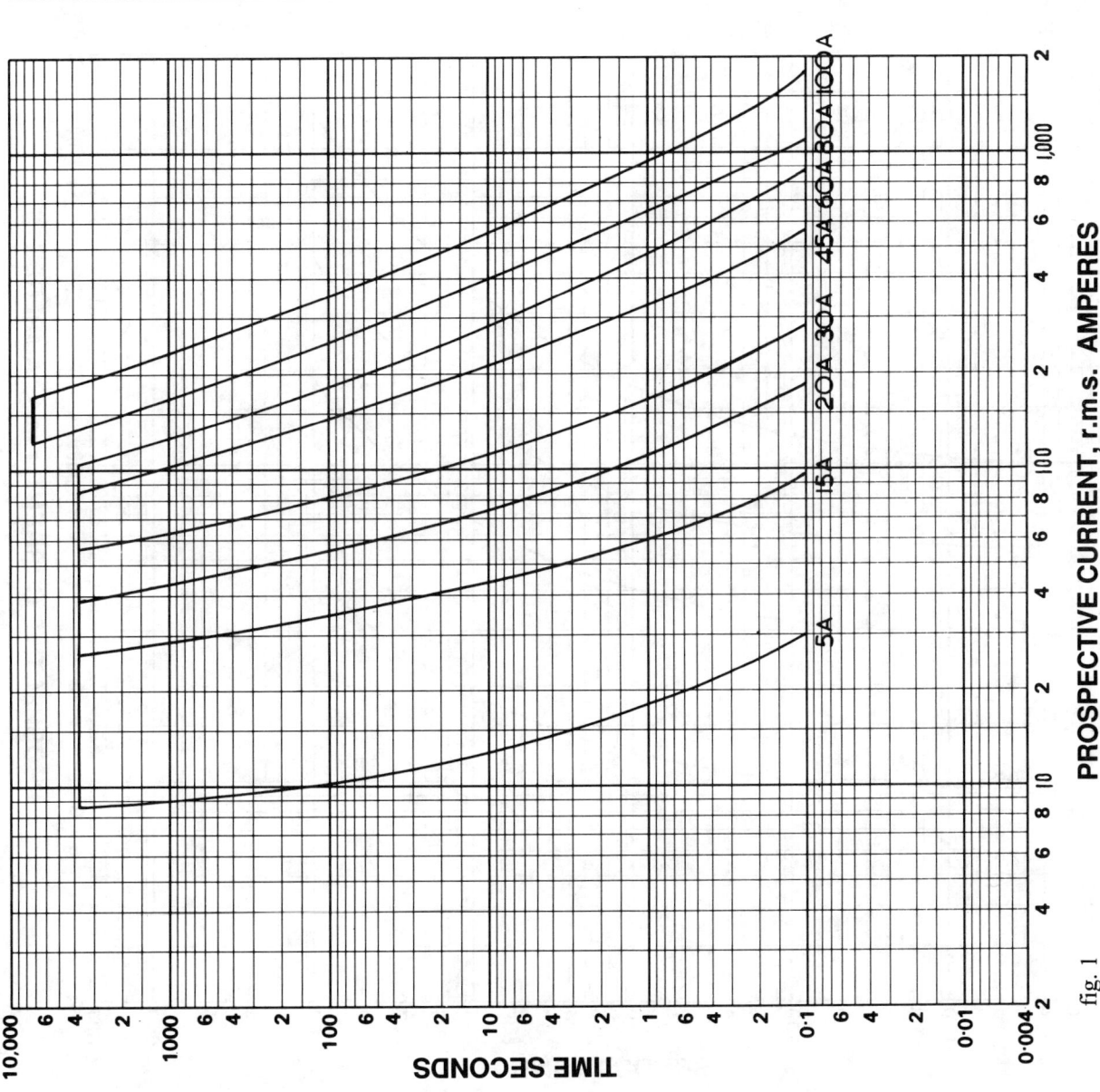

fig. 1

Time/Current characteristics for semi-enclosed fuses to BS 3036

FUSE RATING	CURRENT FOR TIME			
	0.1 sec	0.2 sec	0.4 sec	5 secs
5 A	45 A	32 A	24 A	13 A
15 A	180 A	125 A	90 A	43 A
30 A	450 A	300 A	210 A	87 A
60 A	1300 A	800 A	550 A	205 A

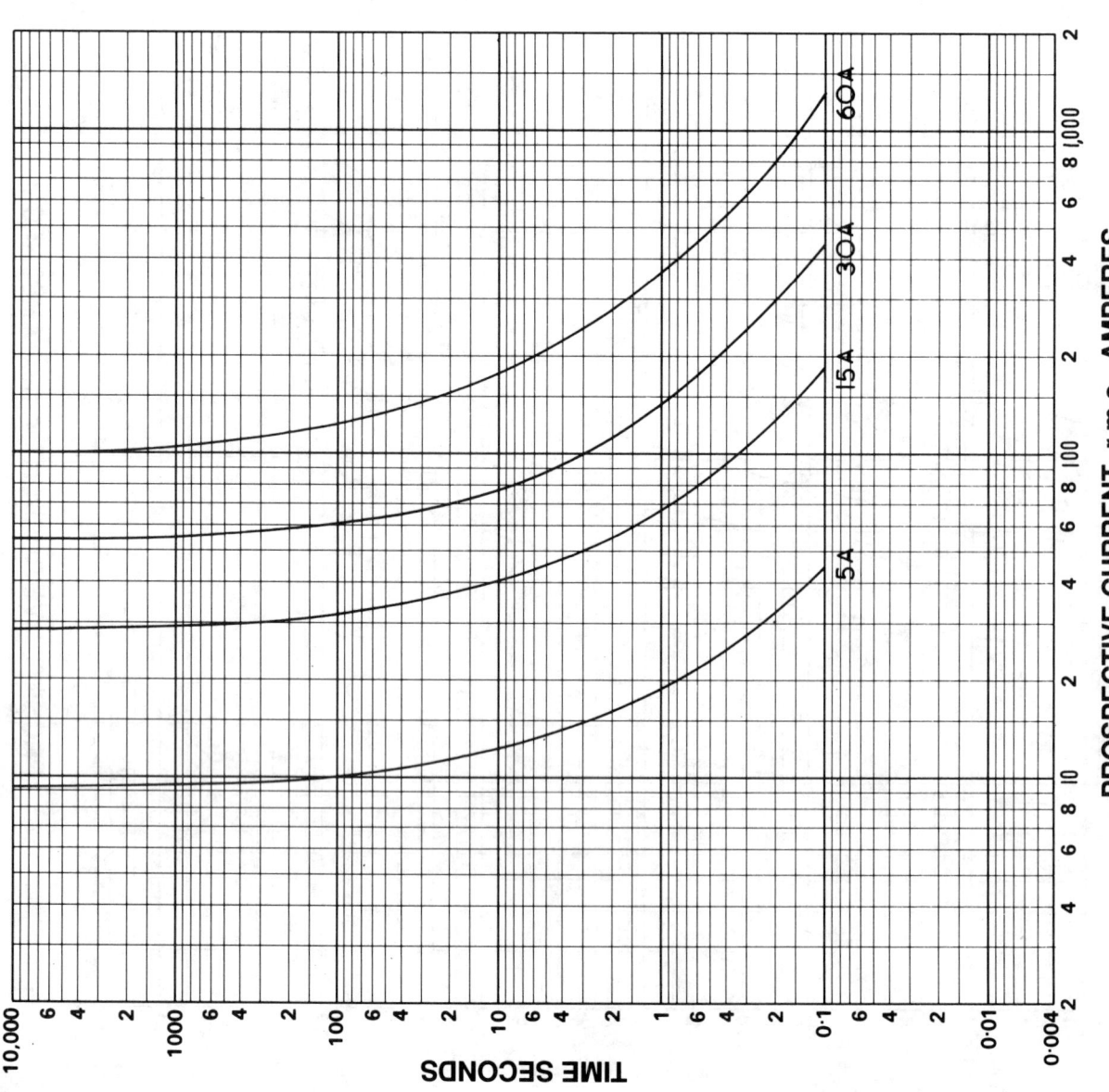

fig. 2A

Time/Current characteristics for semi-enclosed fuses to BS 3036

FUSE RATING	CURRENT FOR TIME			
	0.1 sec	0.2 sec	0.4 sec	5 secs
20 A	260 A	180 A	130 A	60 A
45 A	900 A	580 A	390 A	145 A
100 A	2800 A	1800 A	1200 A	430 A

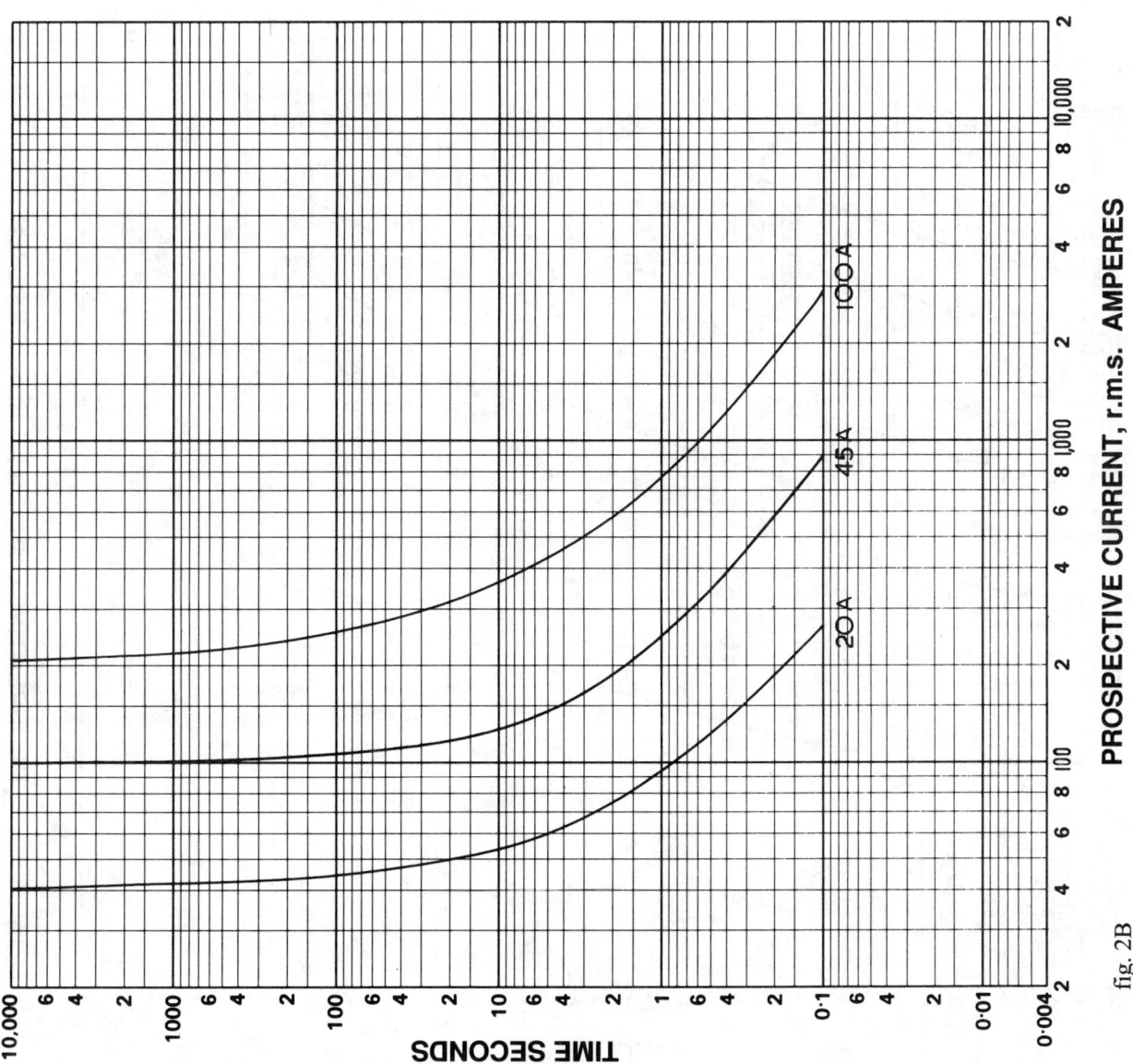

fig. 2B

Time/Current characteristics for fuses to BS 88: Part 2 and Part 6

FUSE RATING	CURRENT FOR TIME			
	0.1 sec	0.2 sec	0.4 sec	5 secs
6 A	36 A	31 A	27 A	17 A
20 A	175 A	150 A	130 A	79 A
32 A	320 A	260 A	220 A	125 A
50 A	540 A	450 A	380 A	220 A
80 A	1100 A	890 A	740 A	400 A
125 A	1800 A	1500 A	1300 A	690 A
200 A	3000 A	2500 A	2200 A	1200 A

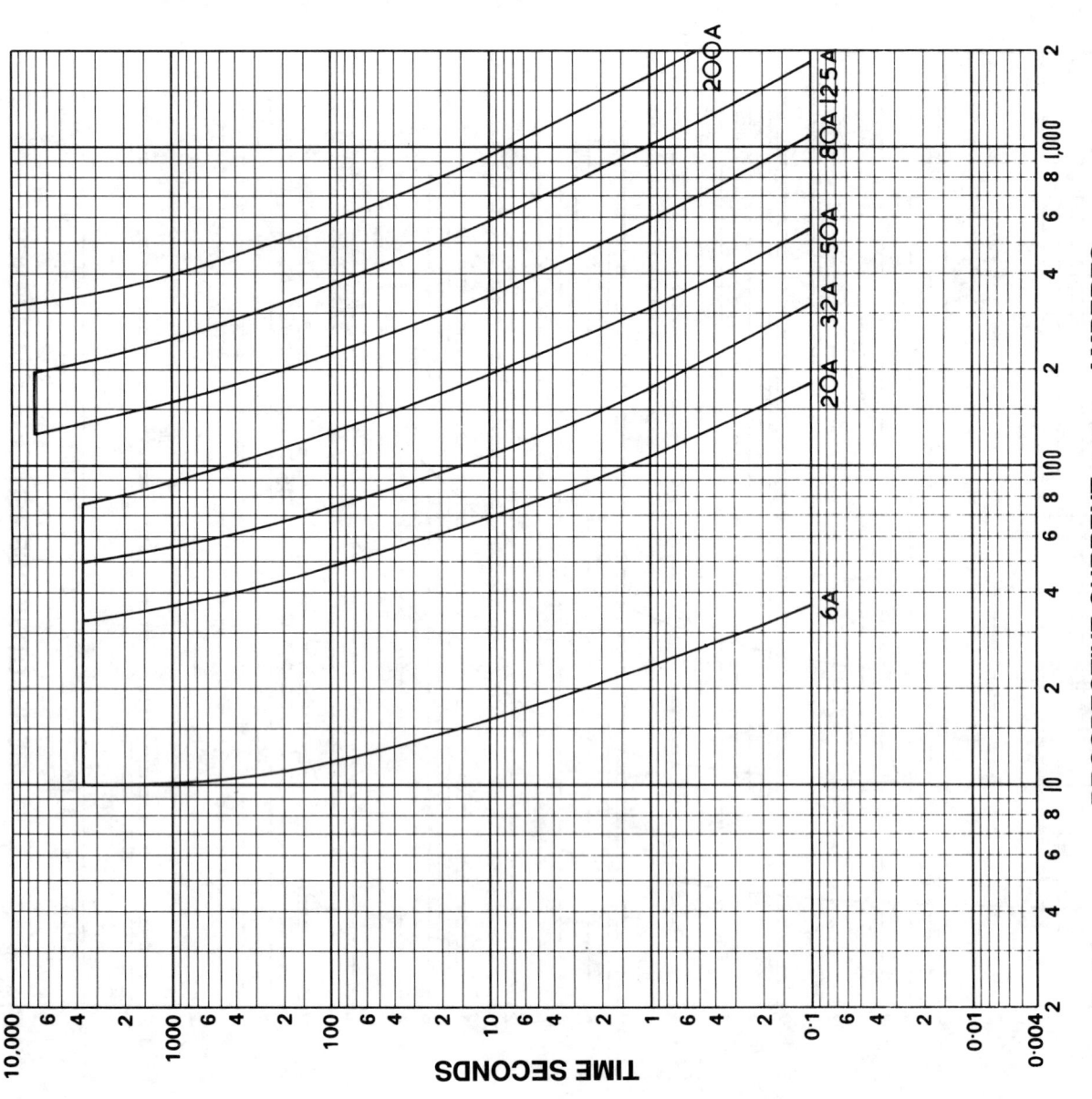

fig. 3A

Time/Current characteristics for fuses to BS 88: Part 2 and Part 6				
FUSE RATING	CURRENT FOR TIME			
	0.1 sec	0.2 sec	0.4 sec	5 secs
10 A	60 A	51 A	45 A	31 A
16 A	120 A	95 A	85 A	55 A
25 A	220 A	180 A	160 A	100 A
40 A	400 A	340 A	280 A	170 A
63 A	710 A	590 A	500 A	280 A
100 A	1400 A	1150 A	980 A	550 A
160 A	2400 A	2000 A	1700 A	900 A

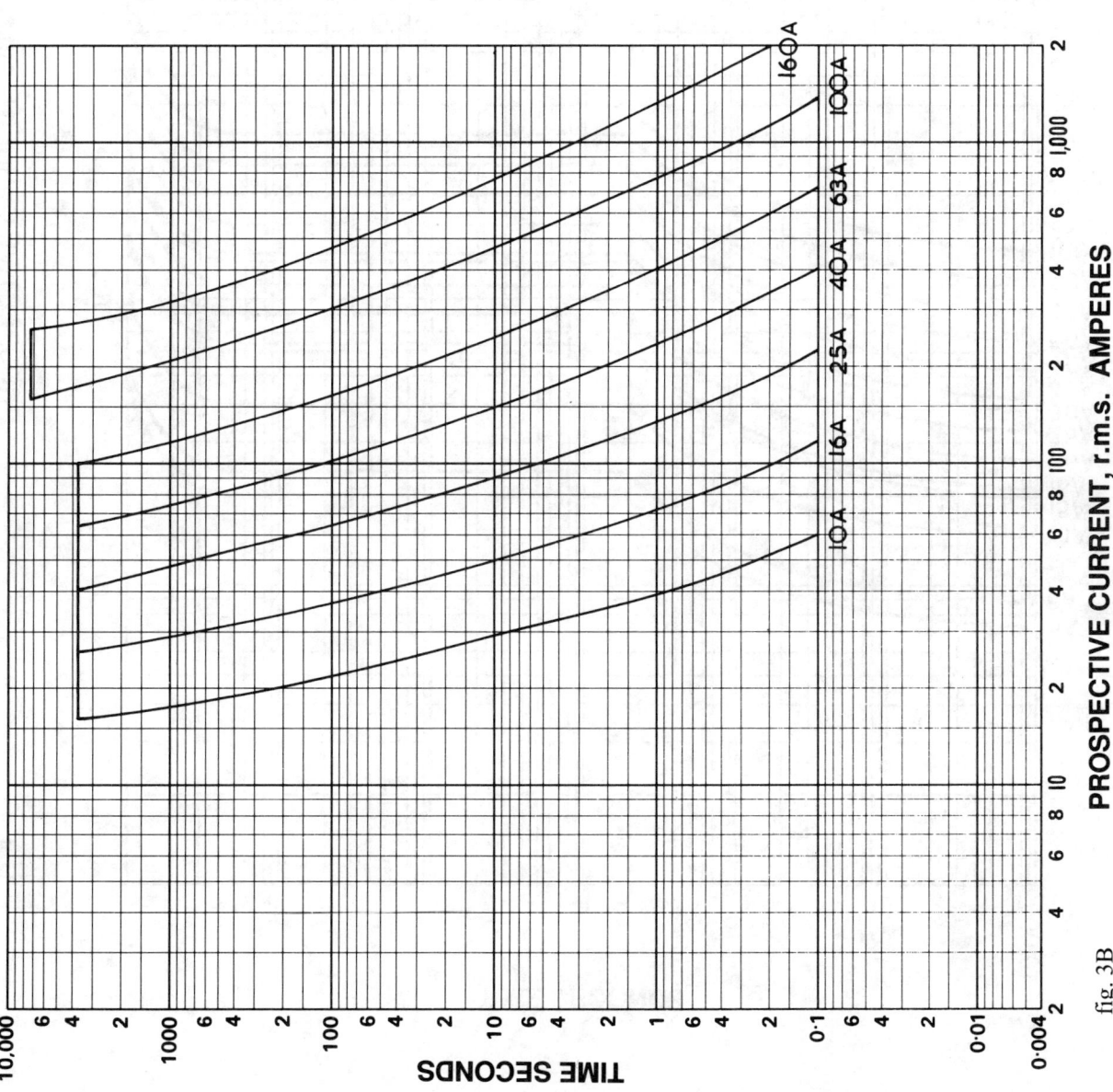

fig. 3B

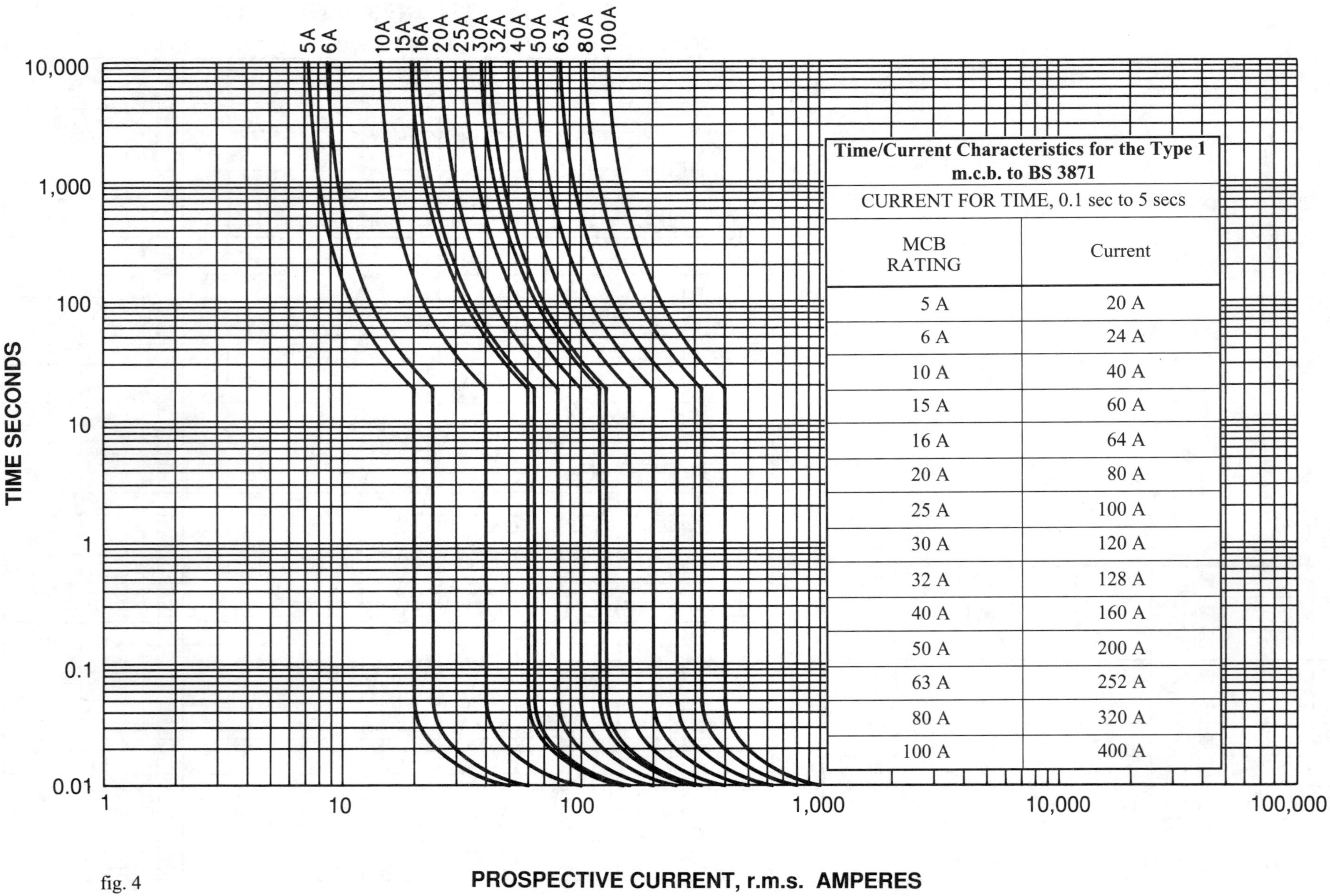

Time/Current Characteristics for the Type 1 m.c.b. to BS 3871

CURRENT FOR TIME, 0.1 sec to 5 secs

MCB RATING	Current
5 A	20 A
6 A	24 A
10 A	40 A
15 A	60 A
16 A	64 A
20 A	80 A
25 A	100 A
30 A	120 A
32 A	128 A
40 A	160 A
50 A	200 A
63 A	252 A
80 A	320 A
100 A	400 A

PROSPECTIVE CURRENT, r.m.s. AMPERES

TIME SECONDS

fig. 4

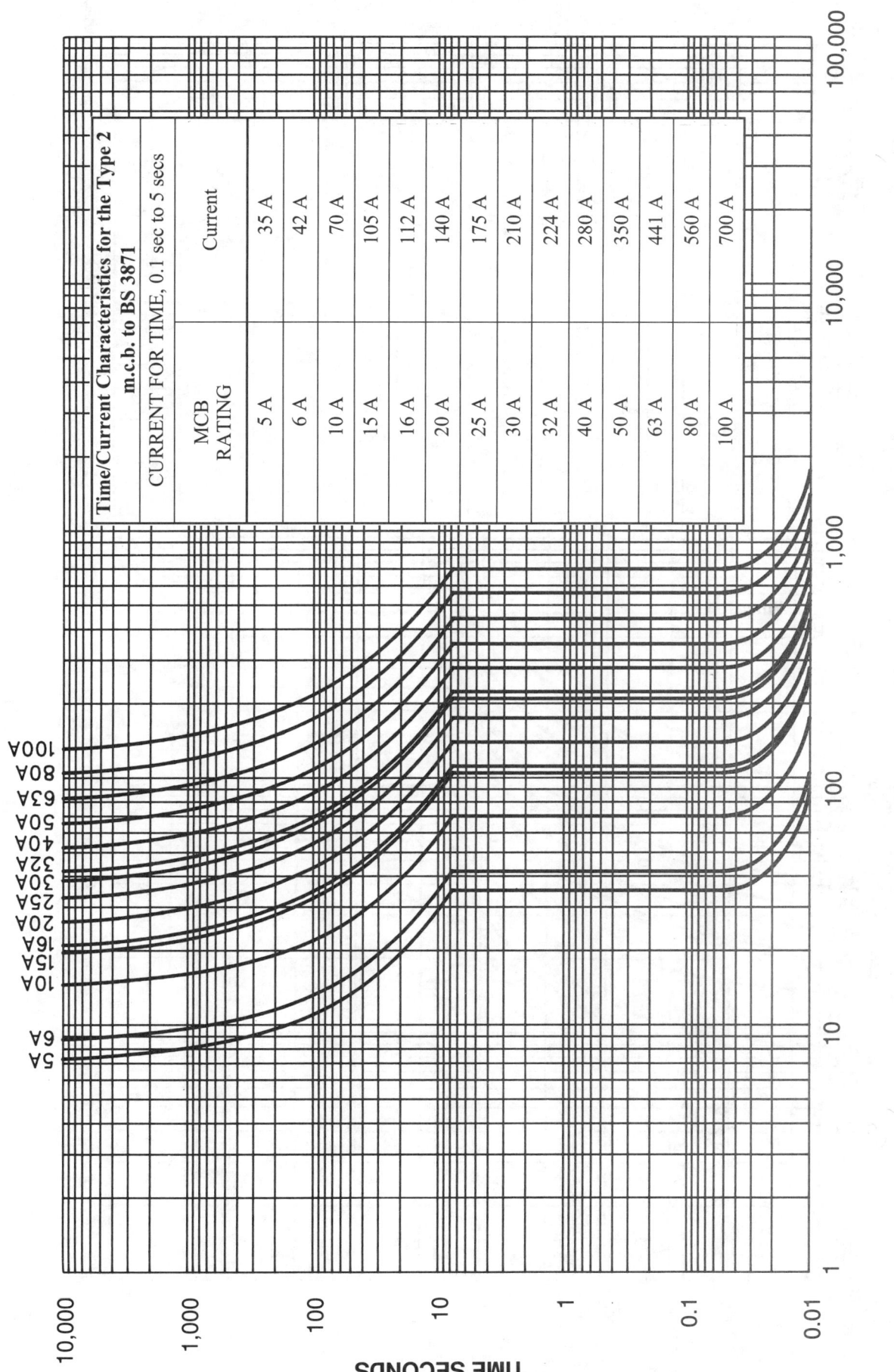

fig. 5

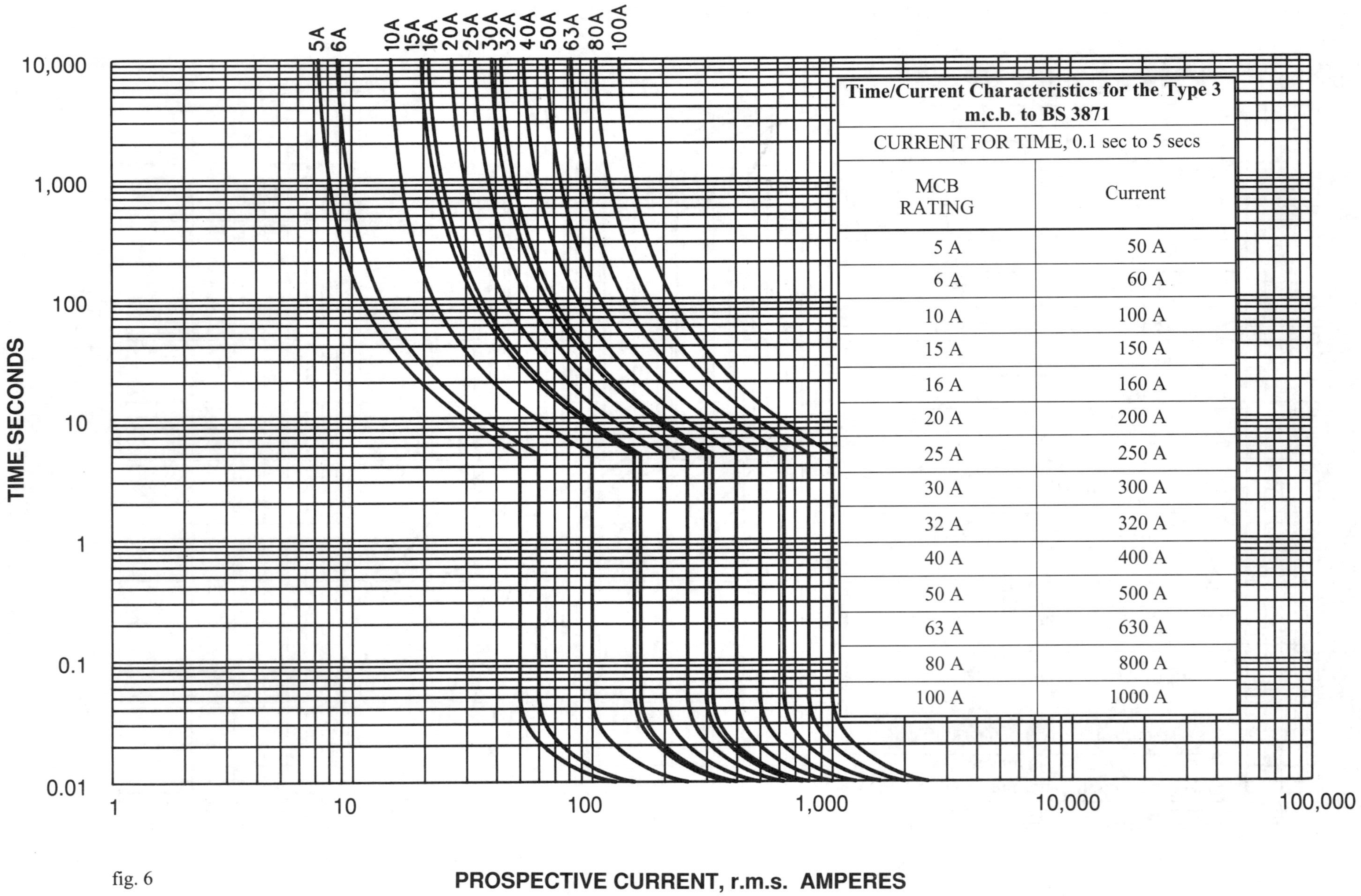

Time/Current Characteristics for the Type 3 m.c.b. to BS 3871

CURRENT FOR TIME, 0.1 sec to 5 secs

MCB RATING	Current
5 A	50 A
6 A	60 A
10 A	100 A
15 A	150 A
16 A	160 A
20 A	200 A
25 A	250 A
30 A	300 A
32 A	320 A
40 A	400 A
50 A	500 A
63 A	630 A
80 A	800 A
100 A	1000 A

fig. 6

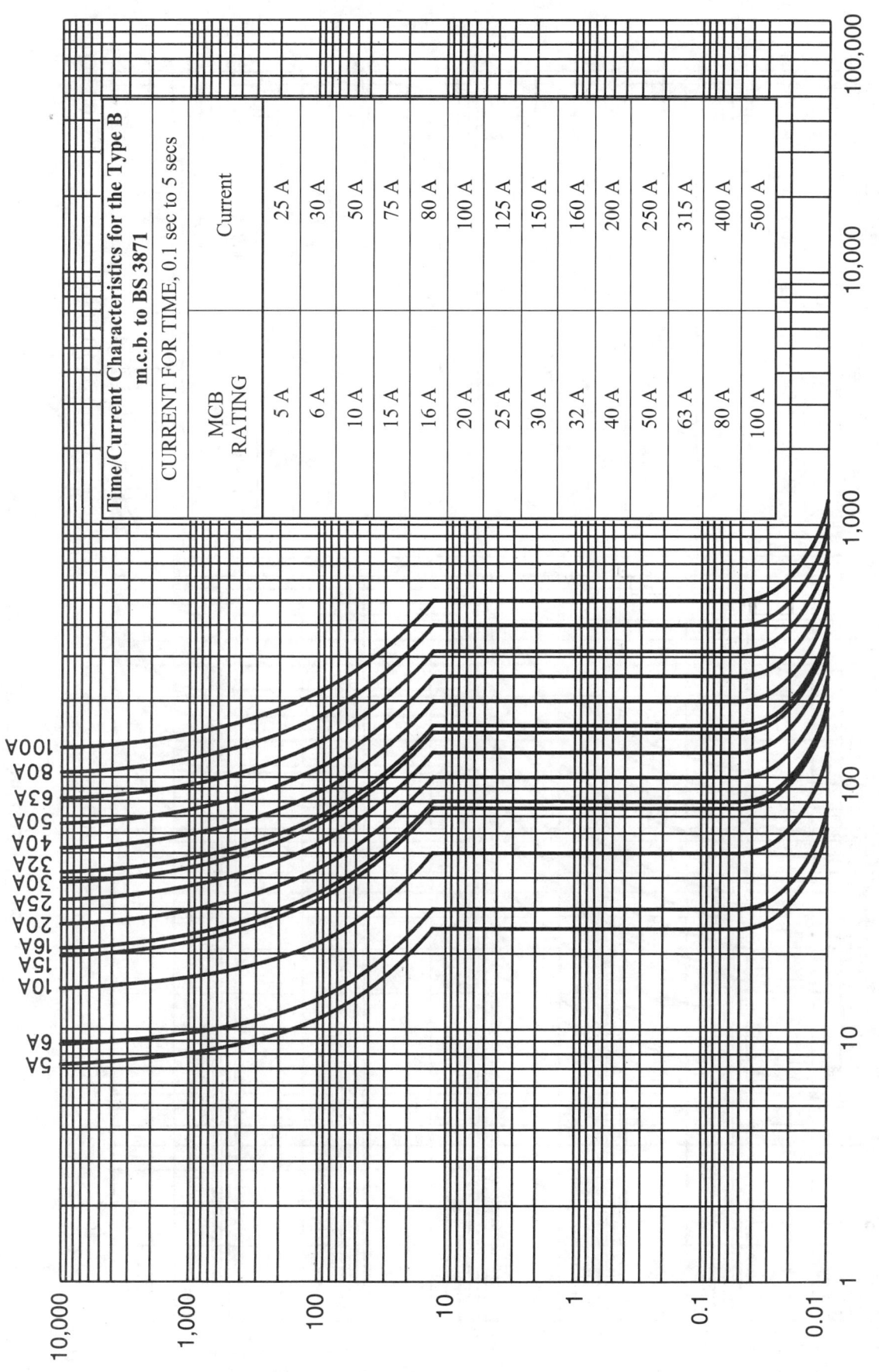

fig. 7

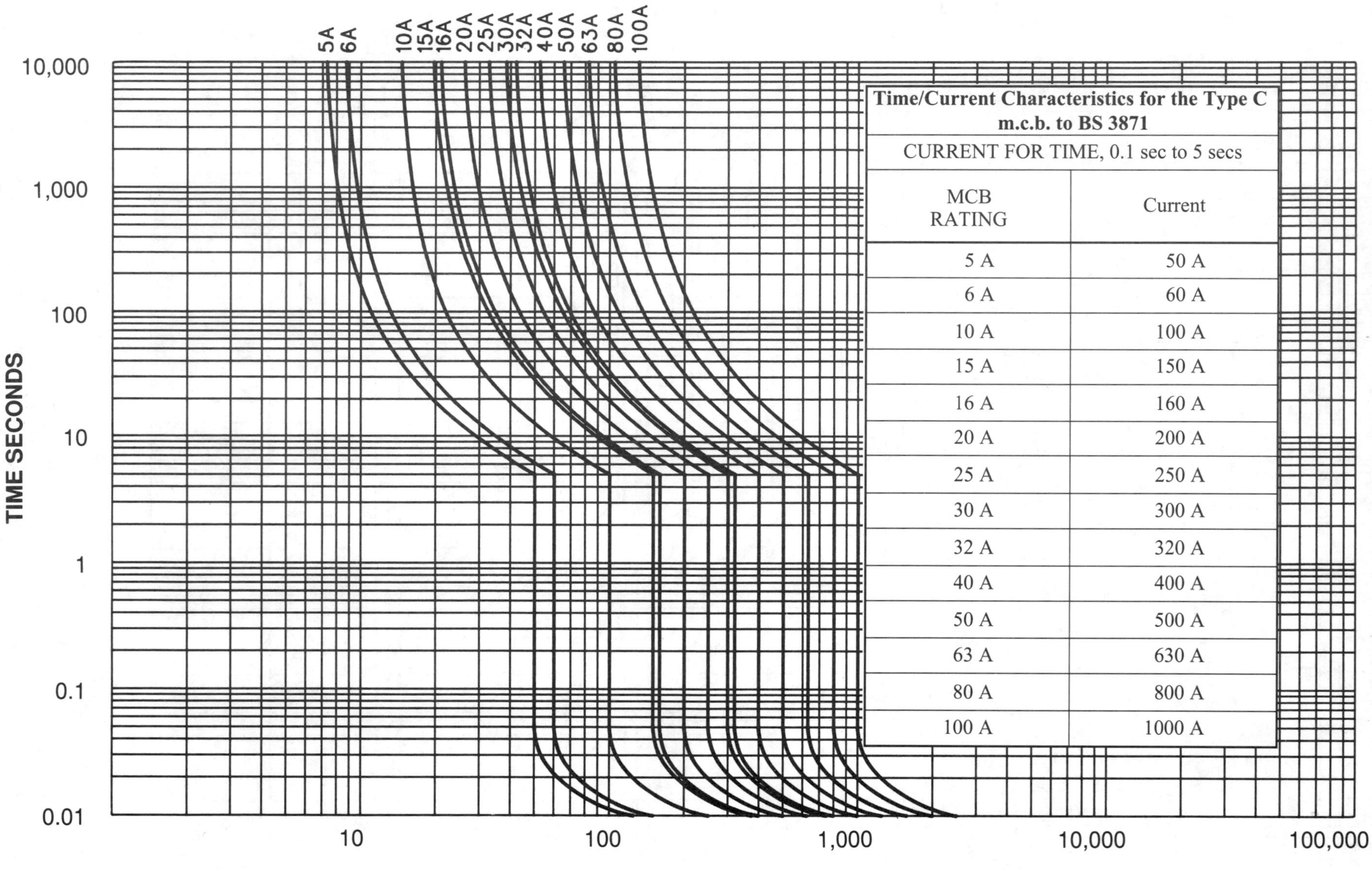

| Time/Current Characteristics for the Type C m.c.b. to BS 3871 ||
| CURRENT FOR TIME, 0.1 sec to 5 secs ||
MCB RATING	Current
5 A	50 A
6 A	60 A
10 A	100 A
15 A	150 A
16 A	160 A
20 A	200 A
25 A	250 A
30 A	300 A
32 A	320 A
40 A	400 A
50 A	500 A
63 A	630 A
80 A	800 A
100 A	1000 A

PROSPECTIVE CURRENT, r.m.s. AMPERES

TIME SECONDS

fig. 8

APPENDIX 4

CURRENT-CARRYING CAPACITY AND VOLTAGE DROP FOR CABLES AND FLEXIBLE CORDS

CONTENTS

Preface to the tables.

Tables:

4A Schedule of methods of installation of cables

4B1 Correction factors for groups of more than one circuit of single-core cables, or more than one multicore cable

4B2 Correction factors for mineral insulated cables.

4B3 Correction factors for cables in enclosed trenches

4C1 Correction factors for ambient temperature where protection is against short circuit only

4C2 Correction factors for ambient temperature where protection against overload is provided by a BS 3036 fuse.

COPPER CONDUCTORS

P.V.C.-INSULATED CABLES

4D1 Single-core non-armoured, with or without sheath
4D2 Multicore non-armoured
4D3 Single-core armoured (non-magnetic armour)
4D4 Multicore armoured

CABLES HAVING THERMOSETTING INSULATION

4E1 Single-core non-armoured, with or without sheath
4E2 Multicore non-armoured
4E3 Single-core armoured (non-magnetic armour)
4E4 Multicore armoured

85°C RUBBER-INSULATED CABLES

4F1 Single-core non-armoured
4F2 Multicore non-armoured

FLEXIBLE CABLES AND CORDS

4H1 60°C rubber-insulated flexible cables
4H2 85°C and 150°C rubber-insulated flexible cables
4H3 Flexible cords

MINERAL INSULATED CABLES

4J1 Bare and exposed to touch, or having an overall covering of p.v.c. - for Reference Method 1
Bare and exposed to touch, or having an overall covering of p.v.c. - for Reference Method 11
Bare and exposed to touch, or having an overall covering of p.v.c. - for Reference Methods 12 and 13

4J2 Bare and neither exposed to touch nor in contact with combustible materials - for Reference Method 1
Bare and neither exposed to touch nor in contact with combustible materials - for Reference Methods 12 and 13.

ALUMINIUM CONDUCTORS

PVC- INSULATED CABLES

4K1 Single-core non-armoured, with or without sheath
4K2 Multicore non-armoured
4K3 Single-core armoured (non-magnetic armour)
4K4 Multicore armoured

CABLES HAVING THERMOSETTING INSULATION

4L1 Single-core, non-armoured
4L2 Multicore, non-armoured
4L3 Single-core armoured (non-magnetic armour)
4L4 Multicore armoured.

APPENDIX 4
CURRENT-CARRYING CAPACITY AND VOLTAGE DROP FOR CABLES AND FLEXIBLE CORDS

PREFACE TO THE TABLES

1. Basis of tabulated current-carrying capacity

The current-carrying capacity set out in this Appendix takes account of IEC Publication 364-5-523 (1983), so far as the latter is applicable. For types of cable not treated in the IEC Publication (e.g. armoured cables) the current-carrying capacity of this Appendix is based on data provided by ERA Technology Ltd., and the British Cable Makers' Confederation (see also ERA Report 69-30 'current rating standards for distribution cables'*).

The tabulated current-carrying capacity relates to continuous loading and is also known as the 'full thermal current rating' of the cable, corresponding to the conductor operating temperature indicated in the headings to the tables concerned. It is intended to provide for a satisfactory life of conductor and insulation subject to the thermal effects of carrying current for sustained periods in normal service. A cable may be seriously damaged, leading to early failure, or its service life may be significantly reduced, if it is operated for any prolonged period at a temperature higher than the indicated value.

In addition, there are other considerations affecting the choice of the cross-sectional area of a conductor, such as the requirements for protection against electric shock (see Chapter 41), protection against thermal effects (see Chapter 42), overcurrent protection (see Chapter 43 and Item 5 below), voltage drop (see Item 7 below) and the limiting temperatures for terminals of equipment to which the conductors are connected.

The tabulated current-carrying capacity relates to a single circuit in the installation methods shown in Table 4A, in an ambient air temperature of 30°C. The current-carrying capacity given in the tables for a.c. operation apply only to frequencies in the range 49 to 61 Hz. For other conditions appropriate correction factors are to be applied as described below.

The current ratings given for single-core armoured cable are for the condition of armour bonded at both ends and to earth.

*ERA Technology Ltd, Cleeve Road, Leatherhead, Surrey KT22 7SA

2. Correction factors for current-carrying capacity

The current-carrying capacity of a cable for continuous service is affected by ambient temperature, by grouping, by partial or total enclosure in thermal insulating material and, for a.c., by frequency. This appendix provides correction factors in these respects as follows.

2.1 *Ambient temperature*

Tables 4C1 and 4C2 give the correction factor to be applied to the tabulated current-carrying capacity depending upon the actual ambient temperature of the location in which the cable is to be installed.

In practice the ambient air temperatures may be determined by thermometers placed in free air as close as practicable to the position at which the cables are installed or are to be installed, subject to the proviso that the measurements are not to be influenced by the heat arising from the cables; thus if the measurements are made while the cables are loaded, the thermometers should be placed about 0.5 m or ten times the overall diameter of the cable, which ever is the lesser, from the cables, in the horizontal plane, or 150 mm below the lowest of the cables.

Tables 4C1 and 4C2 do not take account of temperature increase, if any, due to solar or other infra-red radiation. Where cables are subject to such radiation, the current-carrying capacity may need to be specially calculated.

2.2 *Grouping*

Tables 4B1, 4B2 and 4B3 give the correction factors to be applied to the tabulated current-carrying capacity where cables or circuits are grouped.

2.3 *Other frequencies*

In extreme cases, notably for large multicore cables, the reduction in current-carrying capacity of cables carrying, for example, balanced 400 Hz a.c. compared with the current-carrying capacity at 50 Hz, may be as much as 50%. For small cables and flexible cords, such as may be used to supply individual tools, the difference in the 50 Hz and the 400 Hz current-carrying capacities may be negligible.

3. Effective current-carrying capacity

The current-carrying capacity of a cable corresponds to the maximum current that can be carried in specified conditions without the conductors exceeding the permissible limit of steady state temperature for the type of insulation concerned.

The values of current tabulated represent the effective current-carrying capacity only where no correction factor is applicable. Otherwise the current-carrying capacity corresponds to the tabulated value multiplied by the appropriate factor or factors for ambient temperature, grouping and thermal insulation, as applicable.

Irrespective of the type of overcurrent protective device associated with the conductors concerned, the ambient temperature correction factors to be used when calculating current-carrying capacity (as opposed to those used when selecting cable sizes) are those given in Table 4C1.

4. Relationship of current-carrying capacity to other circuit parameters

The relevant symbols used in the Regulations are as follows:

I_z the current-carrying capacity of a cable for continuous service, under the particular installation conditions concerned.

I_t the value of current tabulated in this appendix for the type of cable and installation method concerned, for a single circuit in an ambient temperature of 30°C.

I_b The design current of the circuit, i.e. the current intended to be carried by the circuit in normal service.

I_n The nominal current or current setting of the device protecting the circuit against overcurrent.

I_2 The operating current (i.e. the fusing current or tripping current for the conventional operating time) of the device protecting the circuit against overload.

C   A correction factor to be applied where the installation conditions differ from those for which values of current-carrying capacity are tabulated in this appendix. The various correction factors are identified as follows:

C_a for ambient temperature

C_g for grouping

C_i for thermal insulation

C_t for operating temperature of conductor.

In all circumstances I_z must be not less than I_b, and I_n also must be not less than I_b.

Where the overcurrent device is intended to afford protection against overload, I_2 must not exceed 1.45 times I_z and I_n must not exceed I_z (see Item 5 below).

Where the overcurrent device is intended to afford short circuit protection only, I_n can be greater than I_z and I_2 can be greater than 1.45 times I_z. The protective device is then to be selected for compliance with Regulation 434-03-03.

5. Overload protection

Where overload protection is required, the type of protection provided does not affect the current-carrying capacity of a cable for continuous service (I_z) but it may affect the choice of conductor size. The operating conditions of a cable are influenced not only by the limiting conductor temperature for continuous service, but also by the conductor temperature which might be attained during the conventional operating time of the overload protection device, in the event of an overload.

This means that the operating current of the protective device must not exceed 1.45 I_z. Where the protective device is a fuse to BS 88 or BS 1361 or a miniature circuit-breaker to BS 3871, this requirement is satisfied by selecting a value of I_z not less than I_n.

In practice, because of the standard steps in nominal rating of fuses and circuit-breakers, it is often necessary to select a value of I_n exceeding I_b. In that case, because it is also necessary for I_z in turn to be not less than the selected value of I_n, the choice of conductor cross-sectional area may be dictated by the overload conditions and the current-carrying capacity (I_z) of the conductors will not always be fully used.

The size needed for a conductor protected against overload by a BS 3036 semi-enclosed fuse can be obtained by the use of a correction factor, 1.45/2=0.725, which results in the same degree of protection as that afforded by other overload protective devices. This factor is to be applied to the nominal rating of the fuse as a divisor, thus indicating the minimum value of I_t required of the conductor to be protected. In this case also, the choice of conductor size is dictated by the overload conditions and the current-carrying capacity (I_z) of the conductors cannot be fully used.

6. Determination of the size of cable to be used

Having established the design current (I_b) of the circuit under consideration, the appropriate procedure described in Items 6.1 to 6.4 below will enable the designer to determine the size of the cable it will be necessary to use.

As a preliminary step it is useful to identify the length of the cable run and the permissible voltage drop for the equipment being supplied, as this may be an over-riding consideration (see Regulation 525-01 and Item 7 of this appendix). The permissible voltage drop in mV, divided by I_b and by the length of run, will give the value of voltage drop in mV/A/m which can be tolerated. A voltage drop not exceeding that value is identified in the appropriate table and the corresponding cross-sectional area of conductor needed on this account can be read off directly before any other calculations are made.

The conductor size necessary from consideration of the conditions of normal load and overload is then determined. All correction factors affecting I_z (that is, the factors for ambient temperature, grouping and thermal insulation) can, if desired, be applied to the values of I_t as multipliers. This involves a process of trial and error until a cross-sectional area is reached which ensures that I_z is not less than I_b and not less than I_n of any protective device it is intended to select. In any event, if a correction factor for protection by a semi-enclosed fuse is necessary, this has to be applied to I_n as a divisor. It is therefore more convenient to apply all the correction factors to I_n as divisors.

This method is used in items 6.1 to 6.3 and produces a value of current and that value (or the next larger value) can readily be located in the appropriate table of current-carrying capacity and the corresponding cross-sectional area of conductor can be identified directly. It should be noted that the value of I_t appearing against the chosen cross-sectional area is not I_z. It is not necessary to know I_z where the size of conductor is chosen by this method, but if it is desired to identify I_z the value is determined by the method indicated in Item 3 above.

However, this method cannot be used for cables installed in enclosed trenches (installation methods 18, 19 and 20 of Table 4A) because the correction factors given in Table 4B3 are related to conductor cross-sectional areas. For such cables it is therefore necessary to use the process of trial and error described in the third paragraph above, selecting on a trial basis a particular size of cable from, for instance, voltage drop considerations.

6.1 *Where overload protection is afforded by a fuse to BS 88 or BS 1361, or a miniature circuit-breaker to BS 3871.*

6.1.1 *For single circuits*

- DIVIDE the nominal current of the protective device (I_n) by any applicable correction factor for ambient temperature (C_a) given in Table 4C1.

- then further DIVIDE by any applicable correction for thermal insulation (C_i).

The size of cable to be used is to be such that its tabulated current-carrying capacity (I_t) is not less than the value of nominal current of the protective device adjusted as above:

$$I_t \geq \frac{I_n}{C_a C_i} \tag{1}$$

6.1.2 *For groups*

- DIVIDE the nominal current of the protective device (I_n) by the correction factor for grouping (C_g) given in Tables 4B1, 4B2 or 4B3:

$$I_t \geq \frac{I_n}{C_g} \tag{2}$$

Alternatively, it may be selected in accordance with the following formulae, provided that the circuits of the group are not liable to simultaneous overload:

$$I_t \geq \frac{I_b}{C_g} \quad , \text{and} \tag{3}$$

$$I_t \geq \sqrt{I_n^2 + 0.48 I_b^2 \left(\frac{1 - C_g^2}{C_g^2} \right)} \tag{4}$$

The size of cable to be used is to be such that its tabulated single-circuit current-carrying capacity (I_t) is not less than the value of I_t calculated in accordance with formula (2) above or, where formulae (3) and (4) are used not less than the larger of the resulting two values of I_t.

Where correction factors C_a and/or C_i are applicable, they are to be applied as divisors to the value of I_t determined by the above formulae.

6.2 *Where the protective device is a semi-enclosed fuse to BS 3036:*

6.2.1 *For single circuits*

- DIVIDE the nominal current of the fuse (I_n) by any applicable correction factor for ambient temperature (C_a) given in Table 4C2

- then further DIVIDE by any applicable correction factor for thermal insulation, (C_i)

- then further DIVIDE by 0.725.

The size of cable to be used is to be such that its tabulated current-carrying capacity (I_t) is not less the value of nominal current of the fuse adjusted as above:

$$I_t \geq \frac{I_n}{0.725 C_a C_i} \tag{5}$$

6.2.2 *For groups*

- DIVIDE the nominal current of the fuse I_n by 0.725 and by the applicable correction factor for grouping (C_g) given in Table 4B1, 4B2 or 4B3:

$$I_t \geq \frac{I_n}{0.725 C_g} \qquad (6)$$

Alternatively, it may be selected by the following formulae, provided that the circuits of the group are not liable to simultaneous overload:

$$I_t \geq \frac{I_b}{C_g}, \text{ and} \qquad (7)$$

$$I_t \geq \sqrt{1.9 I_n^2 + 0.48 I_b^2 \left(\frac{1 - C_g^2}{C_g^2}\right)} \qquad (8)$$

The size of cable to be used is to be such that its tabulated single-circuit current-carrying capacity (I_t) is not less than the value of I_t calculated in accordance with formula (6) above or, where formulae (7) and (8) are used, not less than the larger of the resulting two values of I_t.

Where correction factors C_a and/or C_i are applicable, they are to be applied as divisors to the value of I_t determined by the above formulae.

6.3 *Where overload protection is not required:*

Where Regulation 473-01-04 applies, and the cable under consideration is not required to be protected against overload, the design current of the circuit (I_b) is to be divided by any applicable correction factors, and the size of the cable to be used is to be such that its tabulated current-carrying capacity (I_t) for the installation method concerned is not less than the value of I_b adjusted as above:

$$I_t \geq \frac{I_b}{C_a C_g C_i} \qquad (9)$$

6.4 *Variation of installation conditions along a cable route*

The procedures in items 6.1 to 6.3 above are based on the assumption that all the conditions necessitating the use of correction factors apply to the same part of the route of the conductors of the circuit. Where various factors apply to different parts of the route, each part may be treated separately, or alternatively only the factor or combination of factors appropriate to the most onerous conditions encountered along the route may be applied to the whole of the route. It is permissible to obtain more precise factors by calculation of the various conductor temperature rises that will occur along such a route, provided that the appropriate limiting temperature of the conductor is nowhere exceeded (see Regulation 523-01).

7. Tables of voltage drop

In the tables, values of voltage drop are given for a current of one ampere for a metre run, i.e. for a distance of 1 m along the route taken by the cables, and represent the result of the voltage drops in all the circuit conductors. The values of voltage drop assume that the conductors are at their maximum permitted normal operating temperatures.

The values in the tables, for a.c. operation, apply only to frequencies in the range 49 to 61 Hz and for single-core armoured cables the tabulated values apply where the armour is bonded to earth at both ends. The values of voltage drop for cables operating at higher frequencies may be substantially greater.

For a given run, to calculate the voltage drop (in mV) the tabulated value for the cable concerned has to be multiplied by the length of the run in metres and by the current the cable is intended to carry, namely the design current of the circuit (I_b) in amperes. For three-phase circuits the tabulated mV/A/m values relate to the line voltage and balanced conditions have been assumed.

For cables having conductors of 16 mm^2 or less cross-sectional area their inductances can be ignored and (mV/A/m)$_r$ values only are tabulated. For cables having conductors greater than 16 mm^2 cross-sectional area the impedance values are given as (mV/A/m)$_z$, together with the resistive component (mV/A/m)$_r$ and the reactive component (mV/A/m)$_x$.

The *direct* use of the tabulated $(mV/A/m)_r$ or $(mV/A/m)_z$ values, as appropriate, may lead to pessimistically high calculated values of voltage drop or, in other words, to unnecessarily low values of permitted circuit lengths.

For example, where the design current of a circuit is significantly less than the effective current-carrying capacity of the cable chosen, the actual voltage drop would be less than the calculated value because the conductor temperature (and hence its resistance) will be less than that on which the tabulated mV/A/m had been based.

As regards power factor in a.c. circuits the use of the tabulated mV/A/m values, (for the larger cable sizes, the tabulated $(mV/A/m)_z$ values) to calculate the voltage drop is strictly correct only when the phase angle of the cable equals that of the load. When the phase angle of the cable does not equal that of the load, the direct use of the tabulated mV/A/m or $(mV/A/m)_z$ values leads to a calculated value of voltage drop higher than the actual value. In some cases it may be advantageous to take account of the load power factor when calculating voltage drop.

Where a more accurate assessment of voltage drop is desirable the following methods may be used.

7.1 *Correction for operating temperature*

For cables having conductors of cross-sectional area 16 mm² or less the design value of mV/A/m is obtained by multiplying the tabulated value by a factor C_t, given by

$$C_t = \frac{230 + t_p - (C_a^2 C_g^2 - I_b^2)\dfrac{(t_p - 30)}{I_t^2}}{230 + t_p} \tag{10}$$

where t_p = maximum permitted normal operating temperature, °C

This equation applies only where the overcurrent protective device is other than a BS 3036 fuse and where the actual ambient temperature is equal to or greater than 30° C

NOTE: For convenience, the above formula is based on the resistance-temperature coefficient of 0.004 per °C at 20° C for both copper and aluminium conductors.

For cables having conductors of cross-sectional area greater than 16 mm² only the resistive component of the voltage drop is affected by the temperature and the factor C_t is therefore applied only to the tabulated value of $(mV/A/m)_r$ and the design value of $(mV/A/m)_z$ is given by the vector sum of $C_t (mV/A/m)_r$ and $(mV/A/m)_x$.

For very large conductor sizes where the resistive component of voltage drop is much less than the corresponding reactive part (i.e. when $x/r \geq 3$) this correction factor need not be considered.

7.2 *Correction for load power factor*

For cables having conductors of cross-sectional area of 16 mm² or less the design value of mV/A/m is obtained approximately by multiplying the tabulated value by the power factor of the load, cos Ø.

For cables having conductors of cross-sectional area greater than 16 mm² the design value of mV/A/m is given approximately by:

cos Ø (tabulated $(mV/A/m)_r$) + sin Ø (tabulated $(mV/A/m)_x$)

For single-core cables in flat formation the tabulated values apply to the outer cables and may under-estimate for the voltage drop between an outer cable and the centre cable for cross-sectional areas above 240 mm² and power factors greater than 0.8.

7.3 *Combined correction for both operating temperature and load power factor*

From items 7.1 and 7.2 above, where it is considered appropriate to correct the tabulated mV/A/m values for both operating temperature and load power factor, the design values of mV/A/m are given by:

for cables having conductors of 16 mm² or less cross-sectional area

C_t cos Ø (tabulated mV/A/m)

for cables having conductors of cross-sectional area greater than 16 mm²

C_t cos Ø (tabulated $(mV/A/m)_r$) + sin Ø (tabulated $(mV/A/m)_x$).

8. Methods of installation of cables

Table 4A lists the methods of installation for which this Appendix provides guidance for the selection of the appropriate cable size. The methods of installation distinguished by bold type are reference methods for which the current-carrying capacities given in Tables 4D1 to 4L4 have been determined. For the other methods, an indication is given of the appropriate reference method having values of current-carrying capacity which can safely be applied.

As stated in Regulation 521-07-01 the use of other methods is not precluded, where specified by a suitably qualified electrical engineer; in that case the evaluation of current-carrying capacity may need to be based on experimental work.

TABLE 4A

Schedule of Methods of Installation of Cables

Installation method		Examples	Appropriate Reference Method for determining current-carrying capacity
Number	Description		
1	2	3	4
Open and clipped direct:			
1	**Sheathed cables clipped direct to or lying on a non-metallic surface.**		Method 1
Cables embedded direct in building materials:			
2	Sheathed cables embedded directly in masonry, brickwork, concrete, plaster or the like (other than thermally insulating materials)		Method 1
In conduit:			
3	Single-core non-sheathed cables in metallic or non-metallic conduit on a wall or ceiling		Method 3
4	Single-core non-sheathed cables in metallic or non-metallic conduit in a thermally insulating wall or above a thermally insulating ceiling, the conduit being in contact with a thermally conductive surface on one side.†		Method 4
5	Multicore cables having non-metallic sheath, in metallic or non-metallic conduit on a wall or ceiling		Method 3

† The wall is assumed to consist of an outer weatherproof skin, thermal insulation and an inner skin of plasterboard or wood-like material having a thermal conductance not less than 10 W/m²K. The conduit is fixed so as to be close to, but not necessarily touching, the inner skin. Heat from the cables is assumed to escape through the inner skin only.

TABLE 4A (continued)

Installation method		Examples	Appropriate Reference Method for determining current-carrying capacity
Number	Description		
1	2	3	4
6	Sheathed cables in conduit in a thermally insulating wall etc. (otherwise as Ref Method 4)		Method 4
7	Cables in conduit embedded in masonry, brickwork, concrete, plaster or the like (other than thermally insulating materials)		Method 3
In trunking:			
8	Cables in trunking on a wall or suspended in the air		Method 3
9	Cables in flush floor trunking		Method 3
10	Single-core cables in skirting trunking		Method 3
On trays:			
11	**Sheathed cables on a perforated cable tray, bunched and unenclosed. A perforated cable tray is considered as a tray in which the holes occupy at least 30% of the surface area**		**Method 11**

TABLE 4A (continued)

Installation method		Examples	Appropriate Reference Method for determining current-carrying capacity
Number	Description		
1	2	3	4
In free air, on cleats, brackets or a ladder			
12	Sheathed single-core cables in free air (any supporting metalwork under the cables occupying less than 10% of the plan area): Two or three cables vertically one above the other, minimum distance between cable surfaces equal to the overall cable diameter (D_e); distance from the wall not less than $0.5D_e$ Two or three cables horizontally, with spacings as above Three cables in trefoil, distance between wall and surface of nearest cable $0.5D_e$ or nearest cables $0.75D_e$		Method 12
13	Sheathed multicore cables on ladder or brackets, separation greater than $2D_e$ Sheathed multicore cables in free air distance between wall and cable surface not less than $0.3D_e$ Any supporting metalwork under the cables occupying less than 10% of the plan area		Method 13
14	Cables suspended from or incorporating a catenary wire		Method 12 or 13 as appropriate
Cables in building voids:			
15	Sheathed cables installed directly in a thermally insulating wall or above a thermally insulating ceiling, the cable being in contact with a thermally conductive surface on one side (otherwise as Ref Method No 4)		Method 4

TABLE 4A (continued)

	Installation method	Examples	Appropriate Reference Method for determining current-carrying capacity
Number	Description		
1	2	3	4
16	Sheathed cables in ducts or voids formed by the building structure, other than thermally insulating materials		Method 4 Where the cable has a diameter D_e and the duct has a diameter not greater than $5D_e$ or a perimeter not greater than $20D_e$ Method 3 Where the duct has either a diameter greater than $5D_e$ or a perimeter greater than $20D_e$ NOTE 1- Where the perimeter is greater than $60D_e$, installation Methods 18 to 20, as appropriate, should be used. NOTE 2 - D_e is the overall cable diameter. For groups of cables D_e is the sum of the cable diameters.
Cable in trenches:			
17	Cables supported on the wall of an open or ventilated trench, with spacings as indicated for Ref Method 12 or 13 as appropriate		Method 12 or 13, as appropriate
18	**Cables in enclosed trench 450 mm wide by 300 mm deep (minimum dimensions) including 100 mm cover**	Two single-core cables with surfaces separated by a minimum of one cable diameter; three single-core cables in trefoil and touching throughout. Multicore cables or groups of single-core cables with surfaces minimum of 50 mm	**Method 18** **Use rating factors in Table 4B3**

TABLE 4A (continued)

Installation method		Examples	Appropriate Reference Method for determining current-carrying capacity
Number	Description		
1	2	3	4
19	**Cables in enclosed trench 450 mm wide by 600 mm deep (minimum dimensions) including 100 mm cover**	Single-core cables arranged in flat groups of two or three on the vertical trench wall with surfaces separated by one diameter with a minimum distance of 50 mm between groups. Multicore cables installed with surfaces separated by a minimum* of 75 mm. All cables spaced at least 25 mm from the trench wall	**Method 19 Use rating factors in Table 4B3**
20	**Cables in enclosed trench 600 mm wide by 760 mm deep (minimum dimensions) including 100 mm cover**	Single-core cables arranged in groups of two or three in flat formation with the surfaces separated by one diameter or in trefoil formation with cables touching. Groups separated by a minimum* of 50 mm either horizontally or vertically. Multicore cables installed with surfaces separated by a minimum* of 75 mm either horizontally or vertically. All cables spaced at least 25 mm from the trench wall	**Method 20 Use rating factors in Table 4B3**

* Larger spacing to be used where practicable

TABLE 4B1

Correction factors for groups of more than one circuit of single-core cables, or more than one multicore cable (to be applied to the corresponding current-carrying capacity for a single circuit in Tables 4D1 to 4D4, 4E1 to 4E4, 4F1 and 4F2, 4J1, 4K1 to 4K4, 4L1 to 4L4)

Reference method of installation (see Table 4A)		Correction factor (C_g)													
		Number of circuits or multicore cables													
		2	3	4	5	6	7	8	9	10	12	14	16	18	20
Enclosed (Method 3 or 4) or bunched and clipped direct to a non-metallic surface (Method 1)		0.80	0.70	0.65	0.60	0.57	0.54	0.52	0.50	0.48	0.45	0.43	0.41	0.39	0.38
Single layer clipped to a non-metallic surface (Method 1)	Touching	0.85	0.79	0.75	0.73	0.72	0.72	0.71	0.70	–	–	–	–	–	–
	Spaced*	0.94	0.90	0.90	0.90	0.90	0.90	0.90	0.90	0.90	0.90	0.90	0.90	0.90	0.90
Single layer *multicore* on a perforated metal cable tray, vertical or horizontal (Method 11)	Touching	0.86	0.81	0.77	0.75	0.74	0.73	0.73	0.72	0.71	0.70	–	–	–	–
	Spaced* #	0.91	0.89	0.88	0.87	0.87	–	–	–	–	–	–	–	–	–
Single layer *single-core* on a perforated metal cable tray, touching (Method 11)	Horizontal	0.90	0.85	–	–	–	–	–	–	–	–	–	–	–	–
	Vertical	0.85	–	–	–	–	–	–	–	–	–	–	–	–	–
Single layer multicore touching on ladder supports (Method 13)		0.86	0.82	0.80	0.79	0.78	0.78	0.78	0.77	–	–	–	–	–	–

* Spaced by a clearance between adjacent surfaces of at least one cable diameter (D_e). Where the horizontal clearances between adjacent cables exceeds $2D_e$, no correction factor need be applied.
\# Not applicable to Mineral Insulated Cables see Table 4B2.
** When cables having differing conductor operating temperatures are grouped together, the current rating shall be based upon the lowest operating temperature of any cable in the group.

TABLE 4B2

Correction factors for mineral insulated cables installed on perforated tray, (to be applied to the corresponding current-carrying capacity for single circuits for reference method 11 in Table 4J1A)

Tray Orientation	Arrangement of cables	Number of Trays	Number of multicore cables or circuits					
			1	2	3	4	6	9
Horizontal	Multiconductor cables touching	1	1.0	0.90	0.80	0.80	0.75	0.75
Horizontal	Multiconductor cables spaced‡	1	1.0	1.0	1.0	0.95	0.90	–
Vertical	Multiconductor cables touching	1	1.0	0.90	0.80	0.75	0.75	0.70
Vertical	Multiconductor cables spaced‡	1	1.0	0.90	0.90	0.90	0.85	–
Horizontal	Single conductor cables trefoil separated‡‡	1	1.0	1.0	0.95			
Vertical	Single conductor cables trefoil separated‡‡	1	1.0	0.90	0.90			

‡‡ Separated by a clearance between adjacent surfaces of at least two cable diameters ($2D_e$).
‡ Spaced by a clearance between adjacent surfaces of at least one cable diameter (D_e).

Notes to Tables 4B1 and 4B2

1. The factors in the Table are applicable to groups of cables all of one size. The value of current derived from application of the appropriate factors is the maximum current to be carried by any of the cables in the group.
2. If, due to known operating conditions, a cable is expected to carry not more than 30% of its *grouped* rating, it may be ignored for the purpose of obtaining the rating factor for the rest of the group.
   For example, a group of N loaded cables would normally require a group reduction factor of C_g applied to the tabulated I_t. However, if M cables in the group carry loads which are not greater than $0.3\ C_g I_t$ amperes the other cables can be sized by using the group rating factor corresponding to (N-M) cables.
3. When cables having differing conductor operating temperatures are grouped together, the current rating shall be based on the lowest operating temperature of any cable in the group.
4. Where the horizontal clearances between adjacent cables exceeds $2D_e$, no correction factor need be applied.

TABLE 4B3

Correction factors for cables installed in enclosed trenches
(Installation Methods 18, 19 and 20 of Table 4A)*

The correction factors tabulated below relate to the disposition of cables illustrated in items 18 to 20 of Table 4A and are applicable to the current-carrying capacities for Reference Methods 12 or 13 of Table 4A as given in the relevant tables of this appendix.

Correction factors

Conductor cross-sectional area	Installation Method 18				Installation Method 19			Installation Method 20		
	2 single-core cables, or 1 three- or four-core cable	3 single-core cables, or 2 two-core cables	4 single-core cables, or 2 three- or four-core cables	6 single-core cables, 4 two-core cables, or 3 three- or four-core cables	6 single-core cables, 4 two-core cables, or 3 three- or four-core cables	8 single-core cables, or 4 three- or four-core cables	12 single-core cables, 8 two-core cables, or 6 three- or four-core cables	12 single-core cables, 8 two-core cables or 6 three- or four-core cables	18 single-core cables, 12 two-core cables, or 9 three- or four-core cables	24 single-core cables, 16 two-core cables, or 12 three- or four-core cables
	2	3	4	5	6	7	8	9	10	11
mm^2										
4	0.93	0.90	0.87	0.82	0.86	0.83	0.76	0.81	0.74	0.69
6	0.92	0.89	0.86	0.81	0.86	0.82	0.75	0.80	0.73	0.68
10	0.91	0.88	0.85	0.80	0.85	0.80	0.74	0.78	0.72	0.66
16	0.91	0.87	0.84	0.78	0.83	0.78	0.71	0.76	0.70	0.64
25	0.90	0.86	0.82	0.76	0.81	0.76	0.69	0.74	0.67	0.62
35	0.89	0.85	0.81	0.75	0.80	0.74	0.68	0.72	0.66	0.60
50	0.88	0.84	0.79	0.74	0.78	0.73	0.66	0.71	0.64	0.59
70	0.87	0.82	0.78	0.72	0.77	0.72	0.64	0.70	0.62	0.57
95	0.86	0.81	0.76	0.70	0.75	0.70	0.63	0.68	0.60	0.55
120	0.85	0.80	0.75	0.69	0.73	0.68	0.61	0.66	0.58	0.53
150	0.84	0.78	0.74	0.67	0.72	0.67	0.59	0.64	0.57	0.51
185	0.83	0.77	0.73	0.65	0.70	0.65	0.58	0.63	0.55	0.49
240	0.82	0.76	0.71	0.63	0.69	0.63	0.56	0.61	0.53	0.48
300	0.81	0.74	0.69	0.62	0.68	0.62	0.54	0.59	0.52	0.46
400	0.80	0.73	0.67	0.59	0.66	0.60	0.52	0.57	0.50	0.44
500	0.78	0.72	0.66	0.58	0.64	0.58	0.51	0.56	0.48	0.43
630	0.77	0.71	0.65	0.56	0.63	0.57	0.49	0.54	0.47	0.41

*When cables having different conductor operating temperatures are grouped together the current rating shall be based on the lowest operating temperature of any cable in the group.

TABLE 4C1

Correction factors for ambient temperature where protection is against short-circuit

NOTE: This table applies where the associated overcurrent protective device is intended to provide short-circuit protection only. Except where the device is a semi-enclosed fuse to BS 3036 the table also applies where the device is intended to provide overload protection.

Type of insulation	Operating temperature	Ambient temperature °C													
		25	30	35	40	45	50	55	60	65	70	75	80	85	90	95
Rubber (flexible cables only)	60°C	1.04	1.0	0.91	0.82	0.71	0.58	0.41	—	—	—	—	—	—	—	—
General purpose p.v.c.	70°C	1.03	1.0	0.94	0.87	0.79	0.71	0.61	0.50	0.35	—	—	—	—	—	—
Paper	80°C	1.02	1.0	0.95	0.89	0.84	0.77	0.71	0.63	0.55	0.45	0.32	—	—	—	—
Rubber	85°C	1.02	1.0	0.95	0.90	0.85	0.80	0.74	0.67	0.60	0.52	0.43	0.30	—	—	—
Heat resisting p.v.c.*	85°C	1.03	1.0	0.97	0.94	0.91	0.87	0.84	0.79	0.71	0.61	0.50	0.35	—	—	—
Thermosetting	90°C	1.02	1.0	0.96	0.91	0.87	0.82	0.76	0.71	0.65	0.58	0.50	0.41	0.29	—	—
Mineral	70°C sheath	1.03	1.0	0.93	0.85	0.77	0.67	0.57	0.45	0.31	—	—	—	—	—	—
	105°C sheath	1.02	1.0	0.96	0.92	0.88	0.84	0.80	0.75	0.70	0.65	0.60	0.54	0.47	0.40	0.32

NOTE:
1. Correction factors for flexible cords and for 85°C or 150°C rubber-insulated flexible cables are given in the relevant table of current-carrying capacity.
2. This table also applies when determining the current-carrying capacity of a cable.

*These factors are applicable only to ratings in columns 2 to 5 of Table 4D1.

TABLE 4C2

Correction factors for ambient temperature where the overload protective device is a semi-enclosed fuse to BS 3036.

Type of insulation	Operating temperature	Ambient temperature °C													
		25	30	35	40	45	50	55	60	65	70	75	80	85	90	95
Rubber (flexible cables only)	60°C	1.04	1.0	0.96	0.91	0.87	0.79	0.56	—	—	—	—	—	—	—	—
General purpose p.v.c.	70°C	1.03	1.0	0.97	0.94	0.91	0.87	0.84	0.69	0.48	—	—	—	—	—	—
Paper	80°C	1.02	1.0	0.97	0.95	0.92	0.90	0.87	0.84	0.76	0.62	0.43	—	—	—	—
Rubber	85°C	1.02	1.0	0.97	0.95	0.93	0.91	0.88	0.86	0.83	0.71	0.58	0.41	—	—	—
Heat resisting p.v.c.*	85°C	1.03	1.0	0.97	0.94	0.91	0.87	0.84	0.80	0.76	0.72	0.68	0.49	—	—	—
Thermosetting	90°C	1.02	1.0	0.98	0.95	0.93	0.91	0.89	0.87	0.85	0.79	0.69	0.56	0.39	—	—
Mineral: Bare and exposed to touch or p.v.c. covered	70°C sheath	1.03	1.0	0.96	0.93	0.89	0.86	0.79	0.62	0.42	—	—	—	—	—	—
Bare and not exposed to touch	105°C sheath	1.02	1.0	0.98	0.96	0.93	0.91	0.89	0.86	0.84	0.82	0.79	0.77	0.64	0.55	0.43

NOTE: Correction factors for flexible cords and for 85°C or 150°C rubber-insulated flexible cables are given in the relevant table of current-carrying capacity.

* These factors are applicable only to ratings in columns 2 to 5 of Table 4D1.

COPPER CONDUCTORS

TABLE 4D1A

Single-core p.v.c.-insulated cables, non-armoured, with or without sheath
(COPPER CONDUCTORS)

BS 6004
BS 6231
BS 6346

Ambient temperature: 30°C
Conductor operating temperature: 70°C

CURRENT-CARRYING CAPACITY (Amperes):

NOTE: WHERE THE CONDUCTOR IS TO BE PROTECTED BY A SEMI-ENCLOSED FUSE TO BS 3036, SEE ITEM 6.2 OF THE PREFACE TO THIS APPENDIX.

The current-carrying capacities in columns 2 to 5 are also applicable to flexible cables to BS 6004 table 1(c) and to 85°C heat resisting p.v.c. cables to BS 6231 where the cables are used in fixed installations.

Conductor cross-sectional area	Reference Method 4 (enclosed in conduit in thermally insulating wall etc.)		Reference Method 3 (enclosed in conduit on a wall or in trunking etc.)		Reference Method 1 (clipped direct)		Reference Method 11 (on a perforated cable tray horizontal or vertical)		Reference Method 12 (free air)		
	2 cables, single-phase a.c. or d.c.	3 or 4 cables three-phase a.c.	2 cables, single-phase a.c. or d.c.	3 or 4 cables three-phase a.c.	2 cables, single-phase a.c. or d.c. flat and touching	3 or 4 cables three-phase a.c. flat and touching or trefoil	2 cables, single-phase a.c. or d.c. flat and touching	3 or 4 cables three phase a.c. flat and touching or trefoil	Horizontal flat spaced 2 cables, single-phase a.c. or d.c. or 3 cables three-phase a.c.	Vertical flat spaced 2 cables, single-phase a.c. or d.c. or 3 cables three-phase a.c.	Trefoil 3 cables trefoil three phase a.c.
1	2	3	4	5	6	7	8	9	10	11	12
mm²	A	A	A	A	A	A	A	A	A	A	A
1	11	10.5	13.5	12	15.5	14	—	—	—	—	—
1.5	14.5	13.5	17.5	15.5	20	18	—	—	—	—	—
2.5	19.5	18	24	21	27	25	—	—	—	—	—
4	26	24	32	28	37	33	—	—	—	—	—
6	34	31	41	36	47	43	—	—	—	—	—
10	46	42	57	50	65	59	—	—	—	—	—
16	61	56	76	68	87	79	—	—	—	—	—
25	80	73	101	89	114	104	126	112	146	130	110
35	99	89	125	110	141	129	156	141	181	162	137
50	119	108	151	134	182	167	191	172	219	197	167
70	151	136	192	171	234	214	246	223	281	254	216
95	182	164	232	207	284	261	300	273	341	311	264
120	210	188	269	239	330	303	349	318	396	362	308
150	240	216	300	262	381	349	404	369	456	419	356
185	273	245	341	296	436	400	463	424	521	480	409
240	320	286	400	346	515	472	549	504	615	569	485
300	367	328	458	394	594	545	635	584	709	659	561
400	—	—	546	467	694	634	732	679	852	795	656
500	—	—	626	533	792	723	835	778	982	920	749
630	—	—	720	611	904	826	953	892	1138	1070	855
800	—	—	—	—	1030	943	1086	1020	1265	1188	971
1000	—	—	—	—	1154	1058	1216	1149	1420	1337	1079

TABLE 4D1B

VOLTAGE DROP (per ampere per metre): Conductor operating temperature: 70°C

Conductor cross-sectional area	2 cables – d.c.	2 cables – single-phase a.c.							3 or 4 cables – three-phase a.c.								
		Reference Methods 3 & 4 (Enclosed in conduit etc. in or on a wall)			Reference Methods 1 & 11 (Clipped direct or on trays, touching)			Reference Method 12 (Spaced*)			Reference Methods 3 & 4 (Enclosed in conduit etc. in or on a wall)			Reference Methods 1, 11 & 12 (In trefoil)			Reference Methods 1 & 11 (Flat and touching)			Reference Method 12 (Flat spaced*)		
1	2	3			4			5			6			7			8			9		
mm²	mV	mV	x	z	mV	x	z	mV	x	z	mV	x	z	mV	x	z	mV	x	z	mV	x	z
1	44	44			44			44			38			38			38			38		
1.5	29	29			29			29			25			25			25			25		
2.5	18	18			18			18			15			15			15			15		
4	11	11			11			11			9.5			9.5			9.5			9.5		
6	7.3	7.3			7.3			7.3			6.4			6.4			6.4			6.4		
10	4.4	4.4			4.4			4.4			3.8			3.8			3.8			3.8		
16	2.8	2.8			2.8			2.8			2.4			2.4			2.4			2.4		
	r	r	x	z	r	x	z	r	x	z	r	x	z	r	x	z	r	x	z	r	x	z
25	1.75	1.80	0.33	1.80	1.75	0.20	1.75	1.75	0.29	1.80	1.50	0.29	1.55	1.50	0.175	1.50	1.50	0.25	1.55	1.50	0.32	1.55
35	1.25	1.30	0.31	1.30	1.25	0.195	1.25	1.25	0.28	1.30	1.10	0.27	1.10	1.10	0.170	1.10	1.10	0.24	1.10	1.10	0.32	1.15
50	0.93	0.95	0.30	1.00	0.93	0.190	0.95	0.93	0.28	0.97	0.81	0.26	0.85	0.80	0.165	0.82	0.80	0.24	0.84	0.80	0.32	0.86
70	0.63	0.65	0.29	0.72	0.63	0.185	0.66	0.63	0.27	0.69	0.56	0.25	0.61	0.55	0.160	0.57	0.55	0.24	0.60	0.55	0.31	0.63
95	0.46	0.49	0.28	0.56	0.47	0.180	0.50	0.47	0.27	0.54	0.42	0.24	0.48	0.41	0.155	0.43	0.41	0.23	0.47	0.40	0.31	0.51
120	0.36	0.39	0.27	0.47	0.37	0.175	0.41	0.37	0.26	0.45	0.33	0.23	0.41	0.32	0.150	0.36	0.32	0.23	0.40	0.32	0.30	0.44
150	0.29	0.31	0.27	0.41	0.30	0.175	0.34	0.29	0.26	0.39	0.27	0.23	0.36	0.26	0.150	0.30	0.26	0.23	0.34	0.26	0.30	0.40
185	0.23	0.25	0.27	0.37	0.24	0.170	0.29	0.24	0.26	0.35	0.22	0.23	0.32	0.21	0.145	0.26	0.21	0.22	0.31	0.21	0.30	0.36
240	0.180	0.195	0.26	0.33	0.185	0.165	0.25	0.185	0.25	0.31	0.17	0.23	0.29	0.160	0.145	0.22	0.160	0.22	0.27	0.160	0.29	0.34
300	0.145	0.160	0.26	0.31	0.150	0.165	0.22	0.150	0.25	0.29	0.14	0.23	0.27	0.130	0.140	0.190	0.130	0.22	0.25	0.130	0.29	0.32
400	0.105	0.130	0.26	0.29	0.120	0.160	0.20	0.115	0.25	0.27	0.12	0.22	0.25	0.105	0.140	0.175	0.105	0.21	0.24	0.100	0.29	0.31
500	0.086	0.110	0.26	0.28	0.098	0.155	0.185	0.093	0.24	0.26	0.10	0.22	0.25	0.086	0.135	0.160	0.086	0.21	0.23	0.081	0.29	0.30
630	0.068	0.094	0.25	0.27	0.081	0.155	0.175	0.076	0.24	0.25	0.08	0.22	0.24	0.072	0.135	0.150	0.072	0.21	0.22	0.066	0.28	0.29
800	0.053				0.068	0.150	0.165	0.061	0.24	0.25				0.060	0.130	0.145	0.060	0.21	0.22	0.053	0.28	0.29
1000	0.042				0.059	0.150	0.160	0.050	0.24	0.24				0.052	0.130	0.140	0.052	0.20	0.21	0.044	0.28	0.28

*NOTE: Spacings larger than those specified in Method 12 (see table 4A) will result in larger voltage drop.

TABLE 4D2A

**Multicore p.v.c.-insulated cables, non-armoured
(COPPER CONDUCTORS)**

BS 6004
BS 6346

Ambient temperature: 30°C
Conductor operating temperature: 70°C

CURRENT-CARRYING CAPACITY (Amperes):

* With or without protective conductor.

Circular conductors are assumed for sizes up to and including 16 mm². Values for larger sizes relate to shaped conductors and may safely be applied to circular conductors.

NOTE: WHERE THE CONDUCTOR IS TO BE PROTECTED BY A SEMI-ENCLOSED FUSE TO BS 3036, SEE ITEM 6.2 OF THE PREFACE TO THIS APPENDIX.

Conductor cross-sectional area	Reference Method 1 (enclosed in an insulated wall, etc.)		Reference Method 3 (enclosed in conduit on a wall or ceiling, or in trunking)		Reference Method 1 (clipped direct)		Reference Method 11 (on a perforated cable tray), or Reference Method 13 (free air)	
	1 two-core cable*, single-phase a.c. or d.c.	1 three-core cable* or 1 four-core cable, three-phase a.c.	1 two-core cable*, single-phase a.c. or d.c.	1 three-core cable* or 1 four-core cable, three-phase a.c.	1 two-core cable* single-phase a.c. or d.c.	1 three-core cable*, or 1 four-core cable, three-phase a.c.	1 two-core cable*, single-phase a.c. or d.c.	1 three-core cable*, or 1 four-core cable, three-phase a.c.
1	2	3	4	5	6	7	8	9
mm²	A	A	A	A	A	A	A	A
1	11	10	13	11.5	15	13.5	17	14.5
1.5	14	13	16.5	15	19.5	17.5	22	18.5
2.5	18.5	17.5	23	20	27	24	30	25
4	25	23	30	27	36	32	40	34
6	32	29	38	34	46	41	51	43
10	43	39	52	46	63	57	70	60
16	57	52	69	62	85	76	94	80
25	75	68	90	80	112	96	119	101
35	92	83	111	99	138	119	148	126
50	110	99	133	118	168	144	180	153
70	139	125	168	149	213	184	232	196
95	167	150	201	179	258	223	282	238
120	192	172	232	206	299	259	328	276
150	219	196	258	225	344	299	379	319
185	248	223	294	255	392	341	434	364
240	291	261	344	297	461	403	514	430
300	334	298	394	339	530	464	593	497
400	—	—	470	402	634	557	715	597

TABLE 4D2B

VOLTAGE DROP (per ampere per metre): Conductor operating temperature: 70°C

Conductor cross-sectional area 1	Two-core cable d.c. 2	Two-core cable single-phase a.c. 3			Three- or four-core cable three-phase a.c. 4		
mm²	mV	r	x	mV / z	r	x	mV / z
1	44			44			38
1.5	29			29			25
2.5	18			18			15
4	11			11			9.5
6	7.3			7.3			6.4
10	4.4			4.4			3.8
16	2.8			2.8			2.4
25	1.75	1.75	0.170	1.75	1.50	0.145	1.50
35	1.25	1.25	0.165	1.25	1.10	0.145	1.10
50	0.93	0.93	0.165	0.94	0.80	0.140	0.81
70	0.63	0.63	0.160	0.65	0.55	0.140	0.57
95	0.46	0.47	0.155	0.50	0.41	0.135	0.43
120	0.36	0.38	0.155	0.41	0.33	0.135	0.35
150	0.29	0.30	0.155	0.34	0.26	0.130	0.29
185	0.23	0.25	0.150	0.29	0.21	0.130	0.25
240	0.180	0.190	0.150	0.24	0.165	0.130	0.21
300	0.145	0.155	0.145	0.21	0.135	0.130	0.185
400	0.105	0.115	0.145	0.185	0.100	0.125	0.160

TABLE 4D3A

Single-core armoured p.v.c.-insulated cables (non-magnetic armour) (COPPER CONDUCTORS)

BS 6346

CURRENT-CARRYING CAPACITY (Amperes):

Ambient temperature: 30°C
Conductor operating temperature: 70°C

Conductor cross-sectional area	Reference Method 1 (clipped direct)		Reference Method 11 (on a perforated cable tray)		Reference Method 12 (free air)						
	2 cables, single-phase a.c. or d.c. flat & touching	3 or 4 cables, three-phase a.c. flat & touching	2 cables, single-phase a.c. flat & touching	3 or 4 cables three-phase a.c. flat & touching	2 cables single-phase a.c.		2 cables d.c.		3 or 4 cables, three-phase a.c.		
					Horizontal flat spaced	Vertical flat spaced	Horizontal, spaced	Vertical, spaced	Horizontal flat spaced	Vertical flat spaced	3 cables trefoil
1	2	3	4	5	6	7	8	9	10	11	12
mm²	A	A	A	A	A	A	A	A	A	A	A
50	193	179	205	189	229	217	229	216	230	212	181
70	245	225	259	238	287	272	294	279	286	263	231
95	296	269	313	285	349	332	357	340	338	313	280
120	342	309	360	327	401	383	415	396	385	357	324
185	447	399	469	422	511	489	548	525	490	456	425
240	525	465	550	492	593	568	648	622	566	528	501
300	594	515	624	547	668	640	748	719	616	578	567
400	687	575	723	618	737	707	885	851	674	632	657
500	763	622	805	673	810	777	1035	997	721	676	731
630	843	669	891	728	893	856	1218	1174	771	723	809
800	919	710	976	777	943	905	1441	1390	824	772	886
1000	975	737	1041	808	1008	967	1685	1627	872	816	945

NOTE: Where the conductor is to be protected by a semi-enclosed fuse to BS 3036, see item 6.2 of the preface to this appendix.

TABLE 4D3B

VOLTAGE DROP (per ampere per metre): Conductor operating temperature: 70°C

Conductor cross-sectional area	2 cables d.c.	2 cables – single-phase a.c.							3 or 4 cables – three-phase a.c.								
		Reference Methods 1 & 11 (Touching)			Reference Method 12 (Spaced*)				Reference Methods 1, 11 and 12 (in trefoil touching)				Reference Methods 1 & 11 (Flat and touching)				Reference Method 12 (Flat spaced*)
1	2	3			4				5				6				7
mm²	mV	mV			mV				mV				mV				mV
		r	x	z	r	x	z		r	x	z	r	x	z	r	x	z
50	0.93	0.93	0.22	0.95	0.92	0.30	0.97		0.80	0.190	0.82	0.79	0.26	0.84	0.79	0.34	0.86
70	0.63	0.64	0.21	0.68	0.66	0.29	0.72		0.56	0.180	0.58	0.57	0.25	0.62	0.59	0.32	0.68
95	0.46	0.48	0.20	0.52	0.51	0.28	0.58		0.42	0.175	0.45	0.44	0.25	0.50	0.47	0.31	0.57
120	0.36	0.39	0.195	0.43	0.42	0.28	0.50		0.33	0.170	0.37	0.36	0.24	0.43	0.40	0.30	0.50
150	0.29	0.31	0.190	0.37	0.34	0.27	0.44		0.27	0.165	0.32	0.30	0.24	0.38	0.34	0.30	0.45
185	0.23	0.26	0.190	0.32	0.29	0.27	0.39		0.22	0.160	0.27	0.25	0.23	0.34	0.29	0.29	0.41
240	0.180	0.20	0.180	0.27	0.23	0.26	0.35		0.175	0.160	0.23	0.20	0.23	0.30	0.24	0.28	0.37
300	0.145	0.160	0.180	0.24	0.190	0.26	0.32		0.140	0.155	0.21	0.165	0.22	0.28	0.20	0.28	0.34
400	0.105	0.140	0.175	0.22	0.180	0.24	0.30		0.120	0.130	0.195	0.160	0.21	0.26	0.21	0.25	0.32
500	0.086	0.120	0.170	0.21	0.165	0.23	0.29		0.105	0.145	0.180	0.145	0.20	0.25	0.190	0.24	0.30
630	0.068	0.105	0.165	0.195	0.150	0.22	0.27		0.091	0.145	0.170	0.135	0.195	0.23	0.175	0.22	0.28
800	0.053	0.095	0.160	0.185	0.145	0.21	0.25		0.082	0.140	0.160	0.125	0.180	0.22	0.170	0.195	0.26
1000	0.042	0.091	0.155	0.180	0.140	0.190	0.24		0.079	0.135	0.155	0.125	0.165	0.21	0.165	0.170	0.24

*NOTE: Spacings larger than those specified in Method 12 (see table 4A) will result in larger voltage drop.

TABLE 4D4A

Multicore armoured p.v.c.-insulated cables (COPPER CONDUCTORS)

BS 6346

CURRENT-CARRYING CAPACITY (Amperes):
Ambient temperature: 30°C
Conductor operating temperature: 70°C

NOTE: Where the conductor is to be protected by a semi-enclosed fuse to BS 3036, see Item 6.2 of the Preface to this Appendix.

Conductor cross-sectional area	Reference Method 1 (clipped direct)		Reference Method 11 (on a perforated horizontal or vertical cable tray), or Reference Method 13 (free air)	
	1 two-core cable, single-phase a.c. or d.c.	1 three- or four-core cable, three-phase a.c.	1 two-core cable, single-phase a.c. or d.c.	1 three- or four-core cable, three-phase a.c.
1	2	3	4	5
mm²	A	A	A	A
1.5	21	18	22	19
2.5	28	25	31	26
4	38	33	41	35
6	49	42	53	45
10	67	58	72	62
16	89	77	97	83
25	118	102	128	110
35	145	125	157	135
50	175	151	190	163
70	222	192	241	207
95	269	231	291	251
120	310	267	336	290
150	356	306	386	332
185	405	348	439	378
240	476	409	516	445
300	547	469	592	510
400	621	540	683	590

TABLE 4D4B

VOLTAGE DROP (per ampere per metre): Conductor operating temperature: 70°C

Conductor cross-sectional area	Two-core cable d.c.	Two-core cable single-phase a.c.			Three-or four-core cable three-phase a.c.		
1	2	3			4		
mm²	mV	mV			mV		
1.5	29	29			25		
2.5	18	18			15		
4	11	11			9.5		
6	7.3	7.3			6.4		
10	4.4	4.4			3.8		
16	2.8	2.8			2.4		
		r	x	z	r	x	z
25	1.75	1.75	0.170	1.75	1.50	0.145	1.50
35	1.25	1.25	0.165	1.25	1.10	0.145	1.10
50	0.93	0.93	0.165	0.94	0.80	0.140	0.81
70	0.63	0.63	0.160	0.65	0.55	0.140	0.57
95	0.46	0.47	0.155	0.50	0.41	0.135	0.43
120	0.36	0.38	0.155	0.41	0.33	0.135	0.35
150	0.29	0.30	0.155	0.34	0.26	0.130	0.29
185	0.23	0.25	0.150	0.29	0.21	0.130	0.25
240	0.180	0.190	0.150	0.24	0.165	0.130	0.21
300	0.145	0.155	0.145	0.21	0.135	0.130	0.185
400	0.105	0.115	0.145	0.185	0.100	0.125	0.160

TABLE 4E1A

Single-core cables having thermosetting insulation, non-armoured, with or without sheath
(COPPER CONDUCTORS)

BS 5467   BS 7211

Ambient temperature: 30°C
Conductor operating temperature: 90°C

CURRENT-CARRYING CAPACITY (Amperes):

NOTE: Where the conductor is to be protected by a semi-enclosed fuse to BS 3036, see item 6.2 of the preface to this appendix.

The current-carrying capacity in columns 2 to 5 are also applicable to flexible cables to BS 7211 table 3(b) where the cables are used in fixed installations.

For cable in rigid p.v.c. conduit the values stated in table 4D1 are applicable (see Regulation 521-05).

Where a conductor operates at a temperature exceeding 70°C it shall be ascertained that the equipment connected to the conductor is suitable for the conductor operating temperature (see Regulation 512-02).

Conductor cross-sectional area	Reference Method 4 (enclosed in conduit in thermally insulating wall etc.)		Reference Method 3 (enclosed in conduit on a wall or in trunking etc.)		Reference Method 1 ('clipped direct')		Reference Method 11 (on a perforated cable tray horizontal or vertical)		Reference Method 12 (free air)		
	2 cables, single-phase a.c. or d.c.	3 or 4 cables three-phase a.c.	2 cables, single-phase a.c. or d.c.	3 or 4 cables three-phase a.c.	2 cables, single-phase a.c. or d.c. flat and touching	3 or 4 cables three-phase a.c. flat and touching or trefoil	2 cables, single-phase a.c. or d.c. flat and touching	3 or 4 cables three-phase a.c. flat and touching or trefoil	Horizontal flat spaced, 2 cables, single-phase a.c. or d.c. or 3 cables three-phase	Vertical flat spaced, 2 cables, single-phase a.c. or d.c. or 3 cables three-phase	Trefoil 3 cables, trefoil three-phase a.c.
1	2	3	4	5	6	7	8	9	10	11	12
mm²	A	A	A	A	A	A	A	A	A	A	A
1	14	13	17	15	19	17.5	–	–	–	–	–
1.5	18	17	22	19	25	23	–	–	–	–	–
2.5	24	23	30	26	34	31	–	–	–	–	–
4	33	30	40	35	46	41	–	–	–	–	–
6	43	39	51	45	59	54	–	–	–	–	–
10	58	53	71	63	81	74	–	–	–	–	–
16	76	70	95	85	109	99	–	–	–	–	–
25	100	91	126	111	143	130	158	140	183	163	138
35	124	111	156	138	176	161	195	176	226	203	171
50	149	135	189	168	228	209	293	215	274	246	209
70	189	170	240	214	293	268	308	279	351	318	270
95	228	205	290	259	355	326	375	341	426	389	330
120	263	235	336	299	413	379	436	398	495	453	385
150	300	270	375	328	476	436	505	461	570	524	445
185	341	306	426	370	545	500	579	530	651	600	511
240	400	358	500	433	644	590	686	630	769	711	606
300	459	410	573	493	743	681	794	730	886	824	701
400	–	–	683	584	868	793	915	849	1065	994	820
500	–	–	783	666	990	904	1044	973	1228	1150	936
630	–	–	900	764	1130	1033	1191	1115	1423	1338	1069
800	–	–	–	–	1288	1179	1358	1275	1581	1485	1214
1000	–	–	–	–	1443	1323	1520	1436	1775	1671	1349

TABLE 4E1B

VOLTAGE DROP (per ampere per metre): Conductor operating temperature: 90°C

Conductor cross-sectional area	2 cables d.c.	2 cables – single-phase a.c.											3 or 4 cables – three-phase a.c.										
		Reference Methods 3 & 4 (Enclosed in conduit etc. in or on a wall)			Reference Methods 1 & 11 (Clipped direct or on trays, touching)			Reference Method 12 (Spaced*)			Reference Methods 3 & 4 (Enclosed in conduit etc. in or on a wall)			Reference Methods 1, 11 & 12 (In trefoil)			Reference Methods 1 & 11 (Flat touching)			Reference Method 12 (Flat spaced*)		
1	2	3			4			5			6			7			8			9		
mm²	mV	mV			mV			mV			mV			mV			mV			mV		
1	46	46			46			46			40			40			40			40		
1.5	31	31			31			31			27			27			27			27		
2.5	19	19			19			19			16			16			16			16		
4	12	12			12			12			10			10			10			10		
6	7.9	7.9			7.9			7.9			6.8			6.8			6.8			6.8		
10	4.7	4.7			4.7			4.7			4.0			4.0			4.0			4.0		
16	2.9	2.9			2.9			2.9			2.5			2.5			2.5			2.5		
		r	x	z	r	x	z	r	x	z	r	x	z	r	x	z	r	x	z	r	x	z
25	1.85	1.85	0.31	1.90	1.85	0.190	1.85	1.85	0.28	1.85	1.60	0.27	1.65	1.60	0.165	1.60	1.60	0.190	1.60	1.60	0.27	1.65
35	1.35	1.35	0.29	1.35	1.35	0.180	1.35	1.35	0.27	1.35	1.15	0.25	1.15	1.15	0.155	1.15	1.15	0.180	1.15	1.15	0.26	1.20
50	0.99	1.00	0.29	1.05	0.99	0.180	1.00	0.99	0.27	1.00	0.87	0.25	0.90	0.86	0.155	0.87	0.86	0.180	0.87	0.86	0.26	0.89
70	0.68	0.70	0.28	0.75	0.68	0.175	0.71	0.68	0.26	0.73	0.60	0.24	0.65	0.59	0.150	0.61	0.59	0.175	0.62	0.59	0.25	0.65
95	0.49	0.51	0.27	0.58	0.49	0.170	0.52	0.49	0.26	0.56	0.44	0.23	0.50	0.43	0.145	0.45	0.43	0.170	0.46	0.43	0.25	0.49
120	0.39	0.41	0.26	0.48	0.39	0.165	0.43	0.39	0.25	0.47	0.35	0.23	0.42	0.34	0.140	0.37	0.34	0.165	0.38	0.34	0.24	0.42
150	0.32	0.33	0.26	0.43	0.32	0.165	0.36	0.32	0.25	0.41	0.29	0.23	0.37	0.28	0.140	0.31	0.28	0.165	0.32	0.28	0.24	0.37
185	0.25	0.27	0.26	0.37	0.26	0.165	0.30	0.25	0.25	0.36	0.23	0.23	0.32	0.22	0.140	0.26	0.22	0.165	0.28	0.22	0.24	0.33
240	0.190	0.21	0.26	0.33	0.20	0.160	0.25	0.195	0.25	0.31	0.185	0.22	0.29	0.170	0.140	0.22	0.170	0.160	0.24	0.170	0.24	0.29
300	0.155	0.175	0.25	0.31	0.160	0.160	0.22	0.155	0.25	0.29	0.150	0.22	0.27	0.140	0.140	0.195	0.135	0.160	0.21	0.135	0.24	0.27
400	0.120	0.140	0.25	0.29	0.130	0.155	0.20	0.125	0.24	0.27	0.125	0.22	0.25	0.110	0.135	0.175	0.110	0.160	0.195	0.110	0.24	0.26
500	0.093	0.120	0.25	0.28	0.105	0.155	0.185	0.098	0.24	0.26	0.100	0.22	0.24	0.088	0.135	0.160	0.085	0.160	0.180	0.085	0.24	0.25
630	0.072	0.100	0.25	0.27	0.086	0.155	0.175	0.078	0.24	0.25	0.088	0.21	0.23	0.074	0.135	0.150	0.071	0.160	0.170	0.068	0.23	0.24
800	0.056	—	—	—	0.072	0.150	0.170	0.064	0.24	0.25	—	—	—	0.062	0.130	0.145	0.059	0.155	0.165	0.055	0.23	0.24
1000	0.045	—	—	—	0.063	0.150	0.165	0.054	0.24	0.24	—	—	—	0.055	0.130	0.140	0.050	0.155	0.165	0.047	0.23	0.24

NOTE: Spacings larger than those specified in Method 12 (see table 4A) will result in larger voltage drop.

TABLE 4E2A

Multicore, non-armoured cable having thermosetting insulation (COPPER CONDUCTORS)

BS 5467

Ambient temperature: 30°C
Conductor operating temperature: 90°C

CURRENT-CARRYING CAPACITY (Amperes):

Conductor cross-sectional area	Reference Method 4 (enclosed in an insulated wall, etc.)		Reference Method 3 (enclosed in conduit on a wall or ceiling, or in trunking)		Reference Method 1 (clipped direct)		Reference Method 11 (on a perforated cable tray), or Reference Method 13 (free air)	
	1 two-core cable single-phase a.c. or d.c.	1 three- or four-core cable three-phase a.c.	1 two-core cable single-phase a.c. or d.c.	1 three- or four-core cable three-phase a.c.	1 two-core cable single-phase a.c. or d.c.	1 three- or four-core cable three-phase a.c.	1 two-core cable single-phase a.c. or d.c.	1 three- or four-core cable three-phase a.c.
1	2	3	4	5	6	7	8	9
mm²	A	A	A	A	A	A	A	A
16	76	68	91	80	107	96	115	100
25	99	89	119	105	138	119	149	127
35	121	109	146	128	171	147	185	158
50	145	130	175	154	209	179	225	192
70	183	164	221	194	269	229	289	246
95	220	197	265	233	328	278	352	298
120	253	227	305	268	382	322	410	346
150	290	259	334	300	441	371	473	399
185	329	295	384	340	506	424	542	456
240	386	346	459	398	599	500	641	538
300	442	396	532	455	693	576	741	621
400	—	—	625	536	803	667	865	741

NOTE:
1. Where the conductor is to be protected by a semi-enclosed fuse to BS 3036, see item 6.2 of the preface to the appendix.
2. Where a conductor operates at a temperature exceeding 70°C it shall be ascertained that the equipment connected to the conductor is suitable for the conductor operating temperature (see Regulation 512-02).

TABLE 4E2B

VOLTAGE DROP (per ampere per metre): Conductor operating temperature: 90°C

Conductor cross-sectional area 1	Two-core cable d.c. 2		Two-core cable single-phase a.c. 3			Three-or four-core cable three-phase a.c. 4		
mm²	mV			mV			mV	
16	2.9			2.9			2.5	
		r	x	z	r	x	z
25	1.85	1.85	0.160	1.90	1.60	0.140	1.65
35	1.35	1.35	0.155	1.35	1.15	0.135	1.15
50	0.98	0.99	0.155	1.00	0.86	0.135	0.87
70	0.67	0.67	0.150	0.69	0.59	0.130	0.60
95	0.49	0.50	0.150	0.52	0.43	0.130	0.45
120	0.39	0.40	0.145	0.42	0.34	0.130	0.37
150	0.31	0.32	0.145	0.35	0.28	0.125	0.30
185	0.25	0.26	0.145	0.29	0.22	0.125	0.26
240	0.195	0.200	0.140	0.24	0.175	0.125	0.21
300	0.155	0.160	0.140	0.21	0.140	0.120	0.185
400	0.120	0.130	0.145	0.195	0.115	0.125	0.170

TABLE 4E3A

Single-core cables having thermosetting insulation (non-magnetic armour)
(COPPER CONDUCTORS)

BS 5467
BS 6724

CURRENT-CARRYING CAPACITY (Amperes):

Ambient temperature: 30°C
Conductor operating temperature: 90°C

Conductor cross-sectional area	Reference Method 1 (clipped direct)		Reference Method 11 (on a perforated cable tray)		Reference Method 12 (free air)						
					2 cables single-phase a.c.		2 cables d.c.		3 or 4 cables, three-phase a.c.		
	2 cables, single-phase a.c. or d.c. flat & touching	3 or 4 cables, three-phase a.c. flat & touching	2 cables, single-phase a.c. or d.c. flat & touching	3 or 4 cables three-phase a.c. flat & touching	Horizontal flat spaced	Vertical flat spaced	Horizontal spaced	Vertical spaced	Horizontal flat spaced	Vertical flat spaced	3 cables trefoil
1	2	3	4	5	6	7	8	9	10	11	12
mm²	A	A	A	A	A	A	A	A	A	A	A
50	237	220	253	232	282	266	284	270	288	266	222
70	303	277	322	293	357	337	356	349	358	331	285
95	367	333	389	352	436	412	446	426	425	393	346
120	425	383	449	405	504	477	519	497	485	449	402
150	488	437	516	462	566	539	600	575	549	510	463
185	557	496	587	524	643	614	688	660	618	574	529
240	656	579	689	612	749	714	815	782	715	666	625
300	755	662	792	700	842	805	943	906	810	755	720
400	853	717	899	767	929	889	1137	1094	848	797	815
500	962	791	1016	851	1032	989	1314	1266	923	871	918
630	1082	861	1146	935	1139	1092	1528	1474	992	940	1027
800	1170	904	1246	987	1204	1155	1809	1744	1042	978	1119
1000	1261	961	1345	1055	1289	1238	2100	2026	1110	1041	1214

NOTE:
1. Where the conductor is to be protected by a semi-enclosed fuse to BS 3036, see Item 6.2 of the preface to this appendix.
2. Where a conductor operates at a temperature exceeding 70°C it shall be ascertained that the equipment connected to the conductor is suitable for the conductor operating temperature (see Regulation 512-02).

TABLE 4E3B

VOLTAGE DROP (per ampere per metre): Conductor operating temperature: 90°C

Conductor cross-sectional area	2 cables d.c.	2 cables – single-phase a.c.							3 or 4 cables – three-phase a.c.						
		Reference Methods 1 & 11 (Touching)			Reference Method 12 (Spaced*)			Reference Methods 1, 11 and 12 (in trefoil touching)			Reference Methods 1 & 11 (Flat touching)			Reference Method 12 (Flat spaced*)		
		3			4			5			6			7		
	mV	mV			mV			mV			mV			mV		
1	2	r	x	z	r	x	z	r	x	z	r	x	z	r	x	z
mm²																
50	0.98	0.99	0.21	1.00	0.98	0.29	1.00	0.86	0.180	0.87	0.84	0.25	0.88	0.84	0.33	0.90
70	0.67	0.68	0.200	0.71	0.69	0.29	0.75	0.59	0.170	0.62	0.60	0.25	0.65	0.62	0.32	0.70
95	0.49	0.51	0.195	0.55	0.53	0.28	0.60	0.44	0.170	0.47	0.46	0.24	0.52	0.49	0.31	0.58
120	0.39	0.41	0.190	0.45	0.43	0.27	0.51	0.35	0.165	0.39	0.38	0.24	0.44	0.41	0.30	0.51
150	0.31	0.33	0.185	0.38	0.36	0.27	0.45	0.29	0.160	0.33	0.31	0.23	0.39	0.34	0.29	0.45
185	0.25	0.27	0.185	0.33	0.30	0.26	0.40	0.23	0.160	0.28	0.26	0.23	0.34	0.29	0.29	0.41
240	0.195	0.21	0.180	0.28	0.24	0.26	0.35	0.180	0.155	0.24	0.21	0.22	0.30	0.24	0.28	0.37
300	0.155	0.170	0.175	0.25	0.195	0.25	0.32	0.145	0.150	0.21	0.170	0.22	0.28	0.20	0.27	0.34
400	0.115	0.145	0.170	0.22	0.180	0.24	0.30	0.125	0.150	0.195	0.160	0.21	0.27	0.20	0.27	0.33
500	0.093	0.125	0.170	0.21	0.165	0.24	0.29	0.105	0.145	0.180	0.145	0.20	0.25	0.190	0.24	0.31
630	0.073	0.105	0.165	0.195	0.150	0.23	0.27	0.092	0.145	0.170	0.135	0.195	0.24	0.175	0.23	0.29
800	0.056	0.090	0.160	0.190	0.145	0.23	0.27	0.086	0.140	0.165	0.130	0.180	0.23	0.175	0.195	0.26
1000	0.045	0.092	0.155	0.180	0.140	0.21	0.25	0.080	0.135	0.155	0.125	0.170	0.21	0.165	0.180	0.24

*NOTE: Spacings larger than those specified in Method 12 (see table 4A) will result in larger voltage drop.

TABLE 4E4A

Multicore armoured cables having thermosetting insulation (COPPER CONDUCTORS)

BS 5467
BS 6724

Ambient temperature: 30°C
Conductor operating temperature: 90°C

CURRENT-CARRYING CAPACITY (Amperes):

Conductor cross-sectional area	Reference Method 1 (clipped direct)		Reference Method 11 (on a perforated horizontal or vertical cable tray) or Reference Method 13 (free air)	
	1 two-core cable, single-phase a.c. or d.c.	1 three- or four-core cable, three-phase a.c.	1 two-core cable, single-phase a.c. or d.c.	1 three- or four-core cable, three-phase a.c.
1	2	3	4	5
mm²	A	A	A	A
1.5	27	23	29	25
2.5	36	31	39	33
4	49	42	52	44
6	62	53	66	56
10	85	73	90	78
16	110	94	115	99
25	146	124	152	131
35	180	154	188	162
50	219	187	228	197
70	279	238	291	251
95	338	289	354	304
120	392	335	410	353
150	451	386	472	406
185	515	441	539	463
240	607	520	636	546
300	698	599	732	628
400	787	673	847	728

NOTE:
1. Where the conductor is to be protected by a semi-enclosed fuse to BS 3036, see item 6.2 of the preface to this appendix.
2. Where a conductor operates at a temperature exceeding 70°C it shall be ascertained that the equipment connected to the conductor is suitable for the conductor operating temperature (see Regulation 512-02).

TABLE 4E4B

VOLTAGE DROP (per ampere per metre):   Conductor operating temperature: 90°C

Conductor cross-sectional area 1	Two-core cable d.c. 2	Two-core cable single-phase a.c. 3			Three-or four-core cable three-phase a.c. 4		
mm²	mV	mV			mV		
1.5	31	31			27		
2.5	19	19			16		
4	12	12			10		
6	7.9	7.9			6.8		
10	4.7	4.7			4.0		
16	2.9	2.9			2.5		
		r	x	z	r	x	z
25	1.85	1.85	0.160	1.90	1.60	0.140	1.65
35	1.35	1.35	0.155	1.35	1.15	0.135	1.15
50	0.98	0.99	0.155	1.00	0.86	0.135	0.87
70	0.67	0.67	0.150	0.69	0.59	0.130	0.60
95	0.49	0.50	0.150	0.52	0.43	0.130	0.45
120	0.39	0.40	0.145	0.42	0.34	0.130	0.37
150	0.31	0.32	0.145	0.35	0.28	0.125	0.30
185	0.25	0.26	0.145	0.29	0.22	0.125	0.26
240	0.195	0.20	0.140	0.24	0.175	0.125	0.21
300	0.155	0.16	0.140	0.21	0.140	0.120	0.185
400	0.120	0.13	0.145	0.195	0.115	0.125	0.170

COPPER CONDUCTORS

TABLE 4F1A

Single-core non-armoured cables having 85°C rubber insulation (COPPER CONDUCTORS)

BS 6007
BS 6883

Ambient temperature: 30°C
Conductor operating temperature: 85°C

CURRENT-CARRYING CAPACITY (Amperes):

Conductor cross-sectional area	Reference Method 3 (enclosed in conduit etc. in or on a wall)		Reference Method 1 (clipped direct)		Reference Method 11 (on a perforated cable tray) Horizontal or Vertical		Reference Method 12 (free air)	
	2 cables single-phase a.c. or d.c.	3 or 4 cables three-phase a.c.	2 cables, single-phase a.c. or d.c. flat and touching	3 or 4 cables three-phase a.c. flat and touching or trefoil	2 cables, single-phase a.c. or d.c. flat and touching	3 or 4 cables, three-phase a.c. flat and touching or trefoil	single-phase a.c. or d.c. or 3 or 4 cables three-phase a.c. flat spaced horizontal or vertical	3 cables trefoil three-phase a.c.
1	2	3	4	5	6	7	8	9
mm²	A	A	A	A	A	A	A	A
1	17	15	19	17.5	—	—	—	—
1.5	22	19.5	25	23	—	—	—	—
2.5	30	27	34	31	—	—	—	—
4	40	36	45	42	—	—	—	—
6	52	46	59	54	—	—	—	—
10	72	63	81	75	—	—	—	—
16	96	85	108	100	—	—	—	—
25	127	112	143	133	153	140	154	134
35	157	138	177	164	189	174	192	167
50	190	167	215	199	229	211	235	204
70	242	213	274	254	293	269	303	262
95	293	258	332	308	356	327	370	320
120	339	298	384	357	412	379	431	373
150	372	334	442	411	475	437	499	432
185	428	379	519	469	542	499	573	495
240	510	443	607	553	639	589	679	587
300	593	506	695	636	735	679	786	680
400	719	602	827	755	860	798	929	799
500	835	689	946	865	989	918	1081	919
630	975	791	1088	996	1143	1062	1263	1060

NOTE:
1. Where the conductor is to be protected by a semi-enclosed fuse to BS 3036, see item 6.2 of the preface to the appendix.
2. Where a conductor operates at a temperature exceeding 70°C it shall be ascertained that the equipment connected to the conductor is suitable for the conductor operating temperature (see Regulation 512-02).

TABLE 4F1B

VOLTAGE DROP (per ampere per metre):   Conductor operating temperature: 85°C

Conductor cross-sectional area	2 cables d.c.	2 cables – single-phase a.c.							3 or 4 cables – three-phase a.c.								
		Reference Method 3 (Enclosed in conduit etc. in or on a wall)			Reference Methods 1 & 11 (Clipped direct or on trays, touching)			Reference Method 12 (Spaced*)			Reference Method 3 (Enclosed in conduit etc. in or on a wall)			Reference Methods 1, 11 & 12 (In trefoil touching)			Reference Methods 1 & 11 (Flat touching)			Reference Method 12 (Flat spaced*)		
1	2	3			4			5			6			7			8			9		
mm²	mV	mV			mV			mV			mV			mV			mV			mV		
1	46	46			46			—			40			40			40			—		
1.5	31	31			31			—			26			26			26			—		
2.5	18	18			18			—			16			16			16			—		
4	12	12			12			—			10			10			10			—		
6	7.7	7.7			7.7				6.7			6.7			6.7					
10	4.6	4.6			4.6				4.0			4.0			4.0					
16	2.9	2.9			2.9				2.5			2.5			2.5					
		r	x	z	r	x	z	r	x	z	r	x	z	r	x	z	r	x	z
25	1.80	1.85	0.32	1.90	1.85	0.20	1.85	1.85	0.29	1.85	1.60	0.28	1.65	1.60	0.175	1.60	1.60	0.25	1.60	1.60	0.32	1.65
35	1.30	1.35	0.31	1.40	1.30	0.195	1.35	1.30	0.28	1.35	1.15	0.27	1.20	1.15	0.170	1.15	1.15	0.24	1.15	1.15	0.32	1.20
50	0.95	1.00	0.30	1.05	0.97	0.190	0.99	0.97	0.28	1.00	0.87	0.26	0.91	0.84	0.165	0.86	0.84	0.24	0.88	0.84	0.32	0.90
70	0.65	0.68	0.29	0.74	0.66	0.185	0.69	0.66	0.27	0.72	0.60	0.25	0.65	0.57	0.160	0.60	0.57	0.24	0.62	0.57	0.31	0.65
95	0.48	0.51	0.28	0.58	0.49	0.180	0.52	0.49	0.27	0.56	0.44	0.25	0.51	0.43	0.155	0.45	0.43	0.23	0.48	0.42	0.31	0.52
120	0.38	0.40	0.27	0.49	0.39	0.175	0.43	0.39	0.26	0.47	0.35	0.24	0.43	0.34	0.155	0.37	0.34	0.23	0.41	0.34	0.30	0.45
150	0.30	0.33	0.27	0.42	0.31	0.175	0.36	0.31	0.26	0.40	0.29	0.24	0.37	0.27	0.150	0.31	0.27	0.23	0.35	0.27	0.30	0.40
185	0.25	0.27	0.27	0.38	0.25	0.170	0.30	0.25	0.26	0.36	0.23	0.23	0.33	0.22	0.150	0.26	0.22	0.22	0.31	0.22	0.30	0.37
240	0.190	0.21	0.26	0.33	0.195	0.165	0.26	0.195	0.25	0.32	0.180	0.23	0.29	0.170	0.145	0.22	0.170	0.22	0.28	0.170	0.30	0.34
300	0.150	0.170	0.26	0.31	0.155	0.165	0.23	0.155	0.25	0.29	0.150	0.23	0.27	0.140	0.140	0.195	0.135	0.22	0.26	0.135	0.29	0.32
400	0.115	0.140	0.26	0.30	0.125	0.160	0.20	0.120	0.25	0.28	0.130	0.22	0.26	0.140	0.140	0.175	0.110	0.21	0.24	0.105	0.29	0.31
500	0.091	0.115	0.26	0.28	0.100	0.155	0.185	0.097	0.24	0.26	0.105	0.22	0.24	0.135	0.135	0.165	0.089	0.21	0.23	0.085	0.29	0.30
630	0.072	0.100	0.25	0.27	0.082	0.155	0.175	0.077	0.24	0.25	0.085	0.22	0.24	0.135	0.135	0.155	0.073	0.21	0.22	0.067	0.28	0.29

NOTE: Spacings larger than those specified in Method 12 (see table 4A) will result in larger voltage drop.

TABLE 4F2A

Multicore, sheathed and non-armoured cables having 85°C rubber insulation (COPPER CONDUCTORS)

BS 6883

Ambient temperature: 30°C
Conductor operating temperature: 85°C

CURRENT-CARRYING CAPACITY (Amperes):

Conductor cross-sectional area	Reference Method 3 (enclosed)		Reference Method 1 (clipped direct)		Reference Method 11 (on a perforated cable tray), or Reference Method 13 (free air)	
	1 two-core cable single-phase a.c. or d.c.	1 three- or four-core cable, three-phase a.c.	1 two-core cable, single-phase a.c. or d.c.	1 three- or four-core cable, three-phase a.c.	1 two-core cable single-phase a.c. or d.c.	1 three- or four-core cable, three-phase a.c.
1	2	3	4	5	6	7
mm²	A	A	A	A	A	A
1	16.5	14.5	18	16	19.5	17.5
1.5	21	18.5	23	20	25	22
2.5	29	25	32	28	34	30
4	38	33	43	37	46	40
6	48	43	55	48	59	52
10	66	58	76	66	81	71
16	87	77	103	88	109	94
25	114	100	136	117	144	123
35	139	122	168	144	177	151
50	167	147	201	174	213	186
70	211	185	256	222	272	237
95	254	222	310	269	329	287
120	292	256	359	312	381	333
150	320	287	413	359	438	383
185	368	326	470	409	499	437
240	439	381	553	482	587	515
300	509	436	636	555	675	593

NOTE: 1. Where the conductor is to be protected by a semi-enclosed fuse to BS 3036, see item 6.2 of the preface to this appendix.
2. Where a conductor operates at a temperature exceeding 70°C it shall be ascertained that the equipment connected to the conductor is suitable for the conductor operating temperature (see Regulation 512-02).

TABLE 4F2B

VOLTAGE DROP (per ampere per metre):   **Conductor operating temperature: 85°C**

Conductor cross-sectional area	Two-core cable d.c.	Two-core cable single-phase a.c.			Three-or four-core cable three-phase a.c.		
1	2	3			4		
mm²	mV	mV			mV		
1	46	46			40		
1.5	31	31			26		
2.5	19	19			16		
4	12	12			10		
6	7.7	7.7			6.7		
10	4.6	4.6			4.0		
16	2.9	2.9			2.5		
		r	x	z	r	x	z
25	1.80	1.85	0.175	1.85	1.60	0.150	1.60
35	1.30	1.30	0.170	1.35	1.15	0.150	1.15
50	0.95	0.97	0.170	0.99	0.84	0.145	0.86
70	0.65	0.66	0.165	0.68	0.58	0.140	0.59
95	0.48	0.49	0.160	0.52	0.43	0.140	0.45
120	0.38	0.39	0.160	0.42	0.34	0.135	0.36
150	0.30	0.31	0.155	0.35	0.27	0.135	0.30
185	0.25	0.25	0.155	0.30	0.22	0.130	0.26
240	0.190	0.195	0.150	0.25	0.170	0.130	0.22
300	0.150	0.155	0.150	0.22	0.135	0.130	0.185

TABLE 4H1A

60°C rubber-insulated flexible cables, other than flexible cords

BS 6007

CURRENT-CARRYING CAPACITY (Amperes):
Ambient temperature: 30°C
Conductor operating temperature: 60°C

NOTES:

1. The current ratings tabulated are for cables in free air but may also be used for cables resting on a surface. If the cable is to be wound on a drum on load the ratings should be reduced in accordance with Note 3 below and for cables which may be covered, Note 4 below.

2. WHERE THE CONDUCTOR IS TO BE PROTECTED BY A SEMI-ENCLOSED FUSE TO BS 3036, SEE ITEM 6.2 OF THE PREFACE TO THIS APPENDIX.

3. *Flexible Cables Wound on Reeling Drums*

   The current ratings of cables used on reeling drums are to be reduced by the following factors:

   a) Radial type drum
      ventilated: 85%
      unventilated: 75%

   b) Ventilated cylindrical type drum
      1 layer of cable: 85%
      2 layers of cable: 65%
      3 layers of cable: 45%
      9 layers of cable: 35%

   A radial type drum is one where spiral layers of cable are accommodated between closely spaced flanges; if fitted with solid flanges the ratings given above should be reduced and the drum is described as non-ventilated and if the flanges have suitable apertures as ventilated.

   A ventilated cylindrical cable drum is one where widely spaced flanges and the drum and end flanges are accommodated between have suitable ventilating apertures.

4. Where cable may be covered over or coiled up whilst on load, or the air movement over the cable restricted, the current rating should be reduced.

   It is not possible to specify the amount of reduction but the table of rating factors for reeling drums can be used as a guide.

Conductor cross-sectional area	Single-phase a.c. or d.c. 1 two-core cable — with or without protective conductor	Three-phase a.c. 1 three-core, four-core or five-core cable	Single-phase a.c. or d.c. 2 single core cables
1	2	3	4
mm²	A	A	A
4	30	26	—
6	39	34	—
10	51	47	—
16	73	63	—
25	97	83	140
35	—	102	175
50	—	124	216
70	—	158	258
95	—	192	—
120	—	222	302
150	—	255	347
185	—	291	394
240	—	343	471
300	—	394	541
400	—	—	644
500	—	—	738
630	—	—	861

TABLE 4H1B

VOLTAGE DROP (per ampere per metre): Conductor operating temperature: 60°C

Conductor cross-sectional area	Two-core cable d.c.	Two-core cable single-phase a.c.			1 three-core, four-core or five-core cable three-phase a.c.			2 single-core cables touching			
								d.c.	Single-phase a.c.*		
1	2	3			4			5	6		
mm²	mV	mV			mV			mV	mV		
4	12	12			10			—	—		
6	7.8	7.8			6.7			—	—		
10	4.6	4.6			4.0			—	—		
16	2.9	2.9			2.5			—	—		
		r	x	z	r	x	z		r	x	z
25	1.80	1.80	0.175	1.85	1.55	0.150	1.55	—	—	—	—
35	—	—	—		1.10	0.150	1.15	1.31	1.31	0.21	1.32
50	—	—	—		0.83	0.145	0.84	0.91	0.91	0.21	0.93
70	—	—	—		0.57	0.140	0.58	0.64	0.64	0.20	0.67
95	—	—	—		0.42	0.135	0.44	0.49	0.49	0.195	0.53
120	—	—	—		0.33	0.135	0.36	0.38	0.38	0.190	0.43
150	—	—	—		0.27	0.130	0.30	0.31	0.31	0.190	0.36
185	—	—	—		0.22	0.130	0.26	0.25	0.25	0.190	0.32
240	—	—	—		0.170	0.130	0.21	0.190	0.195	0.185	0.27
300	—	—	—		0.135	0.125	0.185	0.150	0.155	0.180	0.24
400	—	—	—		—	—	—	0.115	0.120	0.175	0.21
500	—	—	—		—	—	—	0.090	0.099	0.170	0.20
630	—	—	—		—	—	—	0.068	0.079	0.170	0.185

*Larger voltage drop will result if the cables are spaced.

TABLE 4H2A

85°C or 150°C rubber-insulated flexible cables

BS 6007

CURRENT-CARRYING CAPACITY (Amperes):

Ambient temperature: 30°C
Conductor operating temperature: 85°C

Conductor cross-sectional area	d.c. or single-phase a.c. (1 two-core cable, with or without protective conductor)	Three-phase a.c. (1 three-core, four-core or five-core cable)	Single-phase a.c. or d.c. 2 single-core cables touching
1	2	3	4
mm²	A	A	A
4	41	36	—
6	53	47	—
10	73	64	—
16	99	86	—
25	131	114	—
35	—	140	192
50	—	170	240
70	—	216	297
95	—	262	354
120	—	303	414
150	—	348	476
185	—	397	540
240	—	467	645
300	—	537	741
400	—	—	885
500	—	—	1017
630	—	—	1190

NOTES:

1. The current ratings tabulated are for cables in free air but may also be used for cables resting on a surface. If the cable is to be wound on a drum on load the ratings should be reduced in accordance with Note 3 below and for cables which may be covered, Note 4 below.

2. WHERE THE CONDUCTOR IS TO BE PROTECTED BY A SEMI-ENCLOSED FUSE TO BS 3036, SEE ITEM 6.2 OF THE PREFACE TO THIS APPENDIX.

3. *Flexible Cables Wound on Reeling Drums*

    The current ratings of cables used on reeling drums are to be reduced by the following factors:

    a) Radial type drum
        ventilated: 85%
        unventilated: 75%

    b) Ventilated cylindrical type drum
        1 layer of cable: 85%
        2 layers of cable: 65%
        3 layers of cable: 45%
        4 layers of cable: 35%

    A radial type drum is one where spiral layers of cable are accommodated between closely spaced flanges; if fitted with solid flanges the ratings given above should be reduced and the drum is described as non-ventilated and if the flanges have suitable apertures as ventilated.

    A ventilated cylindrical cable drum is one where layers of cable are accommodated between widely spaced flanges and the drum and end flanges have suitable ventilating apertures.

4. Where cable may be covered over or coiled up whilst on load, or the air movement over the cable restricted, the current rating should be reduced.

    It is not possible to specify the amount of reduction but the table of rating factors for reeling drums can be used as a guide.

5. The temperature limits given in Table 54C should be taken into account when it is intended to operate these cables at maximum permissible temperature.

6. Where a conductor operates at a temperature exceeding 70°C it shall be ascertained that the equipment connected to the conductor is suitable for the conductor operating temperature (see Regulation 512-02).

7. For 150°C cables, where the correction factors for ambient temperature are used, the conductor operating temperature may be up to 150°C.

CORRECTION FACTOR FOR AMBIENT TEMPERATURE

85°C rubber-insulated cables:

Ambient Temperature	35°C	40°C	45°C	50°C	55°C	60°C	65°C	70°C	75°C	80°C
Correction factor	0.95	0.90	0.85	0.80	0.74	0.67	0.60	0.52	0.43	0.30

150°C rubber-insulated cables:

Ambient Temperature	35°C to 85°C	90°C	95°C	100°C	105°C	110°C	115°C	120°C	125°C	130°C	135°C	140°C	145°C
Correction factor	1.0	0.96	0.92	0.88	0.83	0.78	0.73	0.68	0.62	0.55	0.48	0.39	0.28

NOTE: BS 6007 does not include rubber-insulated cables above 16 mm² nominal cross-sectional area.

TABLE 4H2B

VOLTAGE DROP (per ampere per metre):   Conductor operating temperature: 85°C

Conductor cross-sectional area	1 two-core or 2 single-core cables d.c.	Two-core cable single-phase a.c.			1 three-core, four-core or five-core cable three-phase a.c.			2 single-core cables Touching Single-phase a.c.*		
1	2	3			4			5		
mm²	mV	mV			mV			mV		
			x	z		x	z		x	z
4	13	13			11					
6	8.4	8.4			7.3					
10	5.0	5.0			4.3					
16	3.1	3.1			2.7					
		r			r			r		
25	2.0	2.00	0.175	2.00	1.70	0.150	1.70	—	—	—
35	1.42		—		1.20	0.150	1.20	1.42	0.21	1.43
50	0.99		—		0.90	0.145	0.91	0.99	0.21	1.01
70	0.70		—		0.61	0.140	0.63	0.70	0.20	0.72
95	0.53		—		0.46	0.135	0.48	0.53	0.195	0.56
120	0.41		—		0.36	0.135	0.39	0.41	0.190	0.46
150	0.33		—		0.29	0.130	0.32	0.33	0.190	0.38
185	0.27		—		0.24	0.130	0.27	0.27	0.190	0.33
240	0.21		—		0.185	0.130	0.22	0.21	0.185	0.28
300	0.165		—		0.145	0.125	0.195	0.170	0.180	0.25
400	0.125		—		—	—	—	0.130	0.175	0.22
500	0.098		—		—	—	—	0.105	0.170	0.20
630	0.073		—		—	—	—	0.084	0.170	0.190

NOTES 1. The voltage drop figures given above are based on a conductor operating temperature of 85°C and are therefore not accurate when the operating temperature is in excess of 85°C. In the case of the 150°C cables with a conductor temperature of 150°C the above resistive values should be increased by a factor of 1.2. (This factor is only applicable to the range of 150°C rubber-insulated cables included in BS 6007 i.e. up to 16 mm² nominal cross-sectional area).

*2 Larger voltage drop will result if the cables are spaced.

COPPER CONDUCTORS

TABLE 4H3A

Flexible cords

BS 6141
BS 6500

CURRENT-CARRYING CAPACITY (Amperes): and MASS SUPPORTABLE:

Conductor cross-sectional area	Current-carrying capacity		Maximum mass supportable by twin flexible cord (see Regulation 522-08-06)
	Single-phase a.c.	Three-phase a.c.	
1	2	3	4
mm²	A	A	kg
0.5	3	3	2
0.75	6	6	3
1	10	10	5
1.25	13	–	5
1.5	16	16	5
2.5	25	20	5
4	32	25	5

Where cable is on a reel see notes to the Table in 4H1A.

CORRECTION FACTOR FOR AMBIENT TEMPERATURE

60°C rubber and p.v.c. cords:

Ambient temperature	35°C	40°C	45°C	50°C	55°C
Correction factor	0.91	0.82	0.71	0.58	0.41

85°C rubber cords having a h.o.f.r. sheath or a heat-resisting p.v.c. sheath and for 85°C heat-resisting p.v.c. cords:

Ambient temperature	35°C to 50°C	55°C	60°C	65°C	70°C
	1.0	0.96	0.83	0.67	0.47

150°C rubber cords:

Ambient temperature	35°C to 120°C	125°C	130°C	135°C	140°C	145°C
Correction factor	1.0	0.96	0.85	0.74	0.60	0.42

Glass fibre cords:

Ambient temperature	35°C to 150°C	155°C	160°C	165°C	170°C	175°C
Correction factor	1.0	0.92	0.82	0.71	0.57	0.40

TABLE 4H3B

VOLTAGE DROP (per ampere per metre): Conductor operating temperature: 60°C*

Conductor cross-sectional area	d.c. or single-phase a.c.	three-phase a.c.
1	2	3
mm²	mV	mV
0.5	93	80
0.75	62	54
1	46	40
1.25	37	-
1.5	32	27
2.5	19	16
4	12	10

*NOTE: The tabulated values above are for 60°C rubber-insulated and p.v.c.-insulated flexible cords and for other types of flexible cords they are to be multiplied by the following factors:

For 85°C rubber or p.v.c.-insulated   1.09
150°C rubber-insulated   1.31
185°C glass fibre   1.43

Reserved

TABLE 4J1A

Mineral Insulated cables bare and exposed to touch (see note 2) or having an overall covering of p.v.c.
(COPPER CONDUCTORS AND SHEATH)

BS 6207

Ambient temperature: 30°C
Sheath operating temperature: 70°C

CURRENT-CARRYING CAPACITY (Amperes):

REFERENCE METHOD 1 (CLIPPED DIRECT)

Conductor cross-sectional area	2 single-core cables or 1 two-core cable, single-phase a.c. or d.c.	3 single-core cables in trefoil or 1 three-core cable, three-phase a.c.	3 single-core cables in flat formation, three-phase a.c.	1 four-core cable 3 cores loaded three-phase a.c.	1 four-core cable all cores loaded	1 seven-core cable all cores loaded	1 twelve-core cable all cores loaded	1 nineteen-core cable all cores loaded
1	2	3	4	5	6	7	8	9
mm²	A	A	A	A	A	A	A	A
Light duty 500 V								
1	18.5	15	17	15	13	10	—	—
1.5	23	19	21	19.5	16.5	13	—	—
2.5	31	26	29	26	22	17.5	—	—
4	40	35	38	—	—	—	—	—
Heavy duty 750 V								
1	19.5	16	18	16.5	14.5	11.5	9.5	8.5
1.5	25	21	23	21	18	14.5	12.0	10.0
2.5	34	28	31	28	25	19.5	16.0	—
4	45	37	41	37	32	26	—	—
6	57	48	52	47	41	—	—	—
10	77	65	70	64	55	—	—	—
16	102	86	92	85	72	—	—	—
25	133	112	120	110	94	—	—	—
35	163	137	147	—	—	—	—	—
50	202	169	181	—	—	—	—	—
70	247	207	221	—	—	—	—	—
95	296	249	264	—	—	—	—	—
120	340	286	303	—	—	—	—	—
150	388	327	346	—	—	—	—	—
185	440	371	392	—	—	—	—	—
240	514	434	457	—	—	—	—	—

NOTES:
1. For single-core cables, the sheaths of the circuit are assumed to be connected together at both ends.
2. For bare cables exposed to touch, the tabulated values should be multiplied by 0.9.
3. Regulation 528-01-06 allows the omission of partitions under certain circumstances.

TABLE 4J1A (continued)

**Mineral Insulated cables bare and exposed to touch (see note 2)
or having an overall covering of p.v.c.
(COPPER CONDUCTORS AND SHEATH)**

BS 6207

CURRENT-CARRYING CAPACITY (Amperes):

Ambient temperature: 30°C
Sheath operating temperature: 70°C

REFERENCE METHOD 11 (ON A PERFORATED CABLE TRAY, HORIZONTAL OR VERTICAL)

Conductor cross-sectional area	2 single-core cables touching	1 two-core cables	1 three-core cable, three-phase a.c.	1 four-core cable 3 loaded three-phase a.c.	1 four-core cable all cores loaded	1 seven-core cable all cores loaded	1 twelve-core cable all cores loaded	1 nineteen-core cable all cores loaded	3 single-core cables three-phase a.c.			
	Single phase a.c. or d.c.								Vertical spaced	Horizontal spaced	Flat touching	Trefoil
1	10	11	12	13	14	15	16	17	18	19	20	21
mm²	A	A	A	A	A	A	A	A	A	A	A	A
Light duty 500 V												
1	18.5	19.5	16.5	16	14	11	—	—	19	22	17	16.5
1.5	24	25	21	21	18	14	—	—	25	28	22	21
2.5	31	33	28	28	24	19	—	—	32	37	29	28
4	42	44	37	—	—	—	—	—	43	48	39	37
Heavy duty 750 V												
1	20	21	17.5	18	16	12	10	9	21	24	19	17.5
1.5	25	26	22	23	20	15.5	13	11	27	30	25	22
2.5	34	36	30	30	27	21	17	—	35	41	32	30
4	45	47	40	40	35	28	—	—	47	53	43	40
6	57	60	51	51	44	—	—	—	59	67	54	51
10	78	82	69	68	59	—	—	—	80	90	73	69
16	104	109	92	89	78	—	—	—	105	119	97	92
25	135	142	120	116	101	—	—	—	135	154	125	120
35	165	174	147	—	—	—	—	—	164	187	153	147
50	204	215	182	—	—	—	—	—	202	230	188	182
70	251	264	223	—	—	—	—	—	246	279	229	223
95	301	317	267	—	—	—	—	—	294	333	275	267
120	346	364	308	—	—	—	—	—	335	382	314	308
150	395	416	352	—	—	—	—	—	380	431	358	352
185	448	472	399	—	—	—	—	—	424	482	405	399
240	524	552	466	—	—	—	—	—	472	537	471	466

NOTES: 1. For single-core cables, the sheaths of the circuit are assumed to be connected together at both ends.
2. For bare cables exposed to touch, the tabulated values should be multiplied by 0.9.
3. Regulation 528-01-06 allows the omission of partitions under certain circumstances.

TABLE 4J1A (continued)

**Mineral Insulated cables bare and exposed to touch (see note 2)
or having an overall covering of p.v.c.
(COPPER CONDUCTORS AND SHEATH)**

BS 6207

CURRENT-CARRYING CAPACITY (Amperes):

Ambient temperature: 30°C
Sheath operating temperature: 70°C

REFERENCE METHODS 12 and 13 (FREE AIR)

Conductor cross-sectional area	2 single-core cables or 1 two-core cable, single-phase a.c. or d.c.	3 single-core cables in trefoil or 1 three-core cable, three-phase a.c.	1 four-core cable 3 cores loaded three-phase a.c.	1 four-core cable all cores loaded	1 seven-core cable all cores loaded	1 twelve-core cable all cores loaded	1 nineteen-core cable all cores loaded	3 single-core cables three-phase a.c.		
								Vertical spaced	Horizontal spaced	Touching
1	22	23	24	25	26	27	28	29	30	31
mm^2	A	A	A	A	A	A	A	A	A	A
Light duty 500 V										
1	19.5	16.5	16	14	11	—	—	20	23	18
1.5	25	21	21	18	14	—	—	26	29	23
2.5	33	28	28	24	19	—	—	34	39	31
4	44	37	—	—	—	—	—	45	51	41
Heavy duty 750 V										
1	21	17.5	18	16	12	10	9	22	25	20
1.5	26	22	23	20	15.5	13	11	28	32	26
2.5	36	30	30	27	21	17	—	37	43	34
4	47	40	40	35	28	—	—	49	56	45
6	60	51	51	44	—	—	—	62	71	57
10	82	69	68	59	—	—	—	84	95	77
16	109	92	89	78	—	—	—	110	125	102
25	142	120	116	101	—	—	—	142	162	132
35	174	147	—	—	—	—	—	173	197	161
50	215	182	—	—	—	—	—	213	242	198
70	264	223	—	—	—	—	—	259	294	241
95	317	267	—	—	—	—	—	309	351	289
120	364	308	—	—	—	—	—	353	402	331
150	416	352	—	—	—	—	—	400	454	377
185	472	399	—	—	—	—	—	446	507	426
240	552	466	—	—	—	—	—	497	565	496

NOTES: 1. For single-core cables, the sheaths of the circuit are assumed to be connected together at both ends.
2. For bare cables exposed to touch, the tabulated values should be multiplied by 0.9.
3. Regulation 528-01-06 allows the omission of partitions under certain circumstances.

TABLE 4J1B

Mineral Insulated cables bare and exposed to touch or having an overall covering of p.v.c.
(COPPER CONDUCTORS AND SHEATH)

BS 6207

VOLTAGE DROP (per ampere per metre) for single-phase operation: Sheath operating temperature: 70°C

Conductor cross-sectional area	Two single-core cables Touching			Multicore cables		
1	2			3		
mm²	mV			mV		
1	42			42		
1.5	28			28		
2.5	17			17		
4	10			10		
6	7			7		
10	4.2			4.2		
16	2.6			2.6		
	r	x	z	r	x	z
25	1.65	0.200	1.65	1.65	0.145	1.65
35	1.20	0.195	1.20	—	—	—
50	0.89	0.185	0.91	—	—	—
70	0.62	0.180	0.64	—	—	—
95	0.46	0.175	0.49	—	—	—
120	0.37	0.170	0.41	—	—	—
150	0.30	0.170	0.34	—	—	—
185	0.25	0.165	0.29	—	—	—
240	0.190	0.160	0.25	—	—	—

TABLE 4J1B (continued)

Mineral Insulated cables bare and exposed to touch or having an overall covering of p.v.c. (COPPER CONDUCTORS AND SHEATH)

BS 6207

VOLTAGE DROP (per ampere per metre) for three-phase operation: Sheath operating temperature: 70°C

Conductor cross-sectional area	Three single-core cables									Multicore cables		
	Trefoil Touching			Flat Formation								
				Touching			Spaced 1 cable diameter apart					
	2			3			4			5		
mm²	mV			mV			mV			mV		
1	36			36			36			36		
1.5	24			24			24			24		
2.5	14			14			14			14		
4	9.1			9.1			9.1			9.1		
6	6.0			6.0			6.0			6.0		
10	3.6			3.6			3.6			3.6		
16	2.3			2.3			2.3			2.3		
	r	x	z	r	x	z	r	x	z	r	x	z
25	1.45	0.170	1.45	1.45	0.25	1.45	1.45	0.32	1.50	1.45	0.125	1.45
35	1.05	0.165	1.05	1.05	0.24	1.10	1.05	0.31	1.10	—	—	—
50	0.78	0.160	0.80	0.79	0.24	0.83	0.82	0.30	0.87	—	—	—
70	0.54	0.155	0.56	0.55	0.23	0.60	0.58	0.30	0.65	—	—	—
95	0.40	0.150	0.43	0.41	0.22	0.47	0.44	0.29	0.53	—	—	—
120	0.32	0.150	0.36	0.33	0.22	0.40	0.36	0.28	0.46	—	—	—
150	0.26	0.145	0.30	0.29	0.21	0.36	0.32	0.27	0.42	—	—	—
185	0.21	0.140	0.26	0.25	0.21	0.32	0.28	0.26	0.39	—	—	—
240	0.165	0.140	0.22	0.21	0.20	0.29	0.26	0.25	0.36	—	—	—

TABLE 4J2A

Mineral Insulated cables bare and neither exposed to touch nor in contact with combustible materials
(COPPER CONDUCTORS AND SHEATH)

BS 6207

CURRENT-CARRYING CAPACITY (Amperes):

Ambient temperature: 30°C
Sheath operating temperature: 105°C

REFERENCE METHOD 1 (CLIPPED DIRECT)

Conductor cross-sectional area	2 single-core cables or 1 two-core cable, single-phase a.c. or d.c.	3 single-core cables in trefoil or 1 three-core cable, three-phase a.c.	3 single-core cables in flat formation, three-phase a.c.	1 four-core cable 3 cores loaded three-phase a.c.	1 four-core cable all cores loaded	1 seven-core cable all cores loaded	1 twelve-core cable all cores loaded	1 nineteen-core cable all cores loaded
1	2	3	4	5	6	7	8	9
mm²	A	A	A	A	A	A	A	A
Light duty 500 V								
1	22	19	21	18.5	16.5	13	—	—
1.5	28	24	27	24	21	16.5	—	—
2.5	38	33	36	33	28	22	—	—
4	51	44	47	—	—	—	—	—
Heavy duty 750 V								
1	24	20	24	20	17.5	14	12	10.5
1.5	31	26	30	26	22	17.5	15.5	13
2.5	42	35	41	35	30	24	20	—
4	55	47	53	46	40	32	—	—
6	70	59	67	58	50	—	—	—
16	127	107	119	103	90	—	—	—
25	166	140	154	134	117	—	—	—
35	203	171	187	—	—	—	—	—
50	251	212	230	—	—	—	—	—
70	307	260	280	—	—	—	—	—
95	369	312	334	—	—	—	—	—
120	424	359	383	—	—	—	—	—
150	485	410	435	—	—	—	—	—
185	550	465	492	—	—	—	—	—
240	643	544	572	—	—	—	—	—

NOTES:
1. For single-core cables, the sheaths of the circuit are assumed to be connected together at both ends.
2. No correction factor for grouping need be applied.
3. Where a conductor operates at a temperature exceeding 70°C it shall be ascertained that the equipment connected to the conductor is suitable for the conductor operating temperature (see Regulation 512-02).

TABLE 4J2A (continued)

Mineral Insulated cables bare and neither exposed to touch nor in contact with combustible material
(COPPER CONDUCTORS AND SHEATH)

BS 6207

CURRENT-CARRYING CAPACITY (Amperes):

Ambient temperature: 30°C
Sheath operating temperature: 105°C

REFERENCE METHODS 12 and 13 (FREE AIR)

Conductor cross-sectional area	2 single-core cables or 1 two-core cable, single-phase a.c. or d.c.	3 single-core cables in trefoil or 1 three-core cable, three-phase a.c.	1 four-core cable 3 cores loaded three-phase a.c.	1 four-core cable all cores loaded	1 seven-core cable all cores loaded	1 twelve-core cable all cores loaded	1 nineteen-core cable all cores loaded	3 single-core cables three-phase a.c.		
								Vertical spaced	Horizontal spaced	Touching
1	10	11	12	13	14	15	16	17	18	19
mm²	A	A	A	A	A	A	A	A	A	A
Light duty 500 V										
1	24	21	20	18	14	—	—	26	29	23
1.5	31	26	26	22	18	—	—	33	37	29
2.5	41	35	35	30	24	—	—	43	49	39
4	54	46	—	—	—	—	—	56	64	51
Heavy duty 750 V										
1	26	22	22	19	15	13	11	28	32	25
1.5	33	28	28	24	19	16.5	14	35	40	32
2.5	45	38	37	32	26	22	—	47	54	43
4	60	50	49	43	34	—	—	61	70	56
6	76	64	63	54	—	—	—	78	89	71
10	104	87	85	73	—	—	—	105	120	96
16	137	115	112	97	—	—	—	137	157	127
25	179	150	146	126	—	—	—	178	204	164
35	220	184	—	—	—	—	—	216	248	200
50	272	228	—	—	—	—	—	266	304	247
70	333	279	—	—	—	—	—	323	370	300
95	400	335	—	—	—	—	—	385	441	359
120	460	385	—	—	—	—	—	441	505	411
150	526	441	—	—	—	—	—	498	565	469
185	596	500	—	—	—	—	—	557	629	530
240	697	584	—	—	—	—	—	624	704	617

NOTES:
1. For single-core cables, the sheaths of the circuit are assumed to be connected together at both ends.
2. No correction factor for grouping need be applied.
3. Where a conductor operates at a temperature exceeding 70°C it shall be ascertained that the equipment connected to the conductor is suitable for the conductor operating temperature (see Regulation 512-02).

TABLE 4J2B

**Mineral insulated cables bare and neither exposed to touch nor in contact with combustible materials
(COPPER CONDUCTORS AND SHEATH)**

BS 6207

VOLTAGE DROP (per ampere per metre) for single-phase operation: Sheath operating temperature: 105°C

Conductor cross-sectional area	Two single-core cables Touching				Multicore cables			
1	2				3			
mm²	mV				mV			
1	47				47			
1.5	31				31			
2.5	19				19			
4	12				12			
6	7.8				7.8			
10	4.7				4.7			
16	3.0				3.0			
	r	x	z		r	x	z	
25	1.85	0.180	1.85		1.85	0.145	1.85	
35	1.35	0.175	1.35		—	—	—	
50	1.00	0.170	1.00		—	—	—	
70	0.69	0.165	0.71		—	—	—	
95	0.51	0.160	0.54		—	—	—	
120	0.41	0.160	0.44		—	—	—	
150	0.33	0.155	0.36		—	—	—	
185	0.27	0.150	0.31		—	—	—	
240	0.21	0.150	0.26		—	—	—	

TABLE 4J2B (continued)

Mineral Insulated cables bare and neither exposed to touch nor in contact with combustible materials
(COPPER CONDUCTORS AND SHEATH)

BS 6207

VOLTAGE DROP (per ampere per metre) for three-phase operation: Sheath operating temperature: 105°C

Conductor cross-sectional area	Three single-core cables							Multicore cables		
	Trefoil Touching			Flat Formation						
				Touching			Spaced 1 cable diameter apart			
1	2			3			4			5
mm²	mV			mV			mV			mV
1	40			40			40			40
1.5	27			27			27			27
2.5	16			16			16			16
4	10			10			10			10
6	6.8			6.8			6.8			6.8
10	4.1			4.1			4.1			4.1
16	2.6			2.6			2.6			2.6
	r	x	z	r	x	z	r	x	z	r	x	z
25	1.60	0.160	1.65	1.60	0.23	1.65	1.60	0.31	1.65	1.60	0.125	1.60
35	1.15	0.155	1.20	1.15	0.23	1.20	1.20	0.30	1.25	—	—	—
50	0.87	0.150	0.88	0.88	0.22	0.91	0.90	0.29	0.95	—	—	—
70	0.60	0.145	0.62	0.61	0.22	0.65	0.63	0.29	0.70	—	—	—
95	0.45	0.140	0.47	0.46	0.21	0.50	0.48	0.28	0.56	—	—	—
120	0.36	0.135	0.38	0.37	0.21	0.42	0.39	0.28	0.48	—	—	—
150	0.29	0.135	0.32	0.31	0.20	0.37	0.34	0.27	0.43	—	—	—
185	0.23	0.130	0.27	0.26	0.20	0.33	0.29	0.26	0.39	—	—	—
240	0.180	0.130	0.22	0.22	0.195	0.29	0.26	0.25	0.36	—	—	—

TABLE 4K1A

Single-core p.v.c.-insulated cables, non-armoured, with or without sheath (ALUMINIUM CONDUCTORS)

BS 6004
BS 6346

Ambient temperature: 30°C
Conductor operating temperature: 70°C

CURRENT-CARRYING CAPACITY (Amperes):

Conductor cross-sectional area	Reference Method 4 (enclosed in conduit in thermally insulating wall etc.)		Reference Method 3 (enclosed in conduit on a wall or in trunking etc.)		Reference Method 1 (clipped direct)		Reference Method 11 (on a perforated cable tray horizontal or vertical)		Reference Method 12 (free air)		
	2 cables, single-phase a.c. or d.c.	3 or 4 cables three-phase a.c.	2 cables, single-phase a.c. or d.c.	3 or 4 cables three-phase a.c.	2 cables, single-phase a.c. or d.c. flat and touching	3 or 4 cables three-phase a.c. flat and touching or trefoil	2 cables, single-phase a.c. or d.c. flat and touching	3 or 4 cables three-phase a.c. flat and touching or trefoil	Horizontal flat spaced — 2 cables, single-phase a.c. or d.c. or 3 cables three-phase a.c.	Vertical flat spaced — 2 cables, single-phase a.c. or d.c. or 3 cables three-phase a.c.	Trefoil — 3 cables trefoil three-phase a.c.
1	2	3	4	5	6	7	8	9	10	11	12
mm²	A	A	A	A	A	A	A	A	A	A	A
50	93	84	118	104	134	123	144	132	163	148	128
70	118	107	150	133	172	159	185	169	210	191	165
95	142	129	181	161	210	194	225	206	256	234	203
120	164	149	210	186	245	226	261	240	298	273	237
150	189	170	234	204	283	261	301	277	344	317	274
185	215	194	266	230	324	299	344	317	394	364	316
240	252	227	312	269	384	354	407	375	466	432	375
300	289	261	358	306	444	410	469	433	538	501	435
380	—	—	413	352	511	472	543	502	625	584	507
480	—	—	477	405	591	546	629	582	726	680	590
600	—	—	545	462	679	626	722	669	837	787	680
740	—	—	—	—	771	709	820	761	956	902	776
960	—	—	—	—	900	823	953	886	1125	1066	907
1200	—	—	—	—	1022	926	1073	999	1293	1229	1026

NOTE: Where the conductor is to be protected by a semi-enclosed fuse to BS 3036, see item 6.2 of the preface to this appendix.

TABLE 4K1B

VOLTAGE DROP (per ampere per metre). Conductor operating temperature: 70°C

Conductor cross-sectional area	2 cables d.c.	2 cables – single-phase a.c.											3 or 4 cables – three-phase a.c.										
		Reference Methods 3 & 4 (Enclosed in conduit etc., in or on a wall)			Reference Methods 1 & 11 (Clipped direct or on trays, touching)			Reference Method 12 (Spaced*)			Reference Methods 3 & 4 (Enclosed in conduit etc., in or on a wall)			Reference Methods 1, 11 & 12 (In trefoil touching)			Reference Methods 1 & 11 (Flat touching)			Reference Method 12 (Flat spaced*)		
	2	3			4			5			6			7			8			9		
mm²	mV	mV				mV			mV			mV			mV			mV			mV		
		r	x	z	r	x	z	r	x	z	r	x	z	r	x	z	r	x	z	r	x	z
50	1.55	1.60	0.30	1.60	1.55	0.190	1.55	1.55	0.28	1.55	1.35	0.26	1.40	1.35	0.165	1.35	1.35	0.24	1.35	1.35	0.32	1.40
70	1.05	1.10	0.30	1.15	1.05	0.185	1.05	1.05	0.27	1.10	0.94	0.26	0.97	0.91	0.160	0.92	0.91	0.24	0.94	0.91	0.31	0.96
95	0.77	0.81	0.29	0.86	0.77	0.185	0.79	0.77	0.27	0.82	0.70	0.25	0.74	0.67	0.160	0.69	0.67	0.23	0.71	0.67	0.31	0.74
120	0.61	0.64	0.29	0.70	0.61	0.180	0.64	0.61	0.27	0.67	0.55	0.25	0.61	0.53	0.155	0.55	0.53	0.23	0.58	0.53	0.31	0.61
150	0.49	0.51	0.28	0.59	0.49	0.175	0.52	0.49	0.26	0.55	0.45	0.24	0.51	0.42	0.155	0.45	0.42	0.23	0.48	0.42	0.30	0.52
185	0.39	0.42	0.28	0.50	0.40	0.175	0.43	0.39	0.26	0.47	0.36	0.24	0.44	0.34	0.150	0.37	0.34	0.23	0.41	0.34	0.30	0.46
240	0.30	0.32	0.27	0.42	0.30	0.170	0.35	0.30	0.26	0.40	0.28	0.24	0.37	0.26	0.150	0.30	0.26	0.22	0.35	0.26	0.30	0.40
300	0.24	0.26	0.27	0.37	0.24	0.170	0.30	0.24	0.26	0.35	0.23	0.23	0.32	0.21	0.145	0.26	0.21	0.22	0.31	0.21	0.30	0.36
380	0.190	0.22	0.27	0.35	0.195	0.165	0.26	0.195	0.25	0.32	0.190	0.23	0.30	0.170	0.145	0.22	0.170	0.22	0.28	0.170	0.29	0.34
480	0.150	0.180	0.26	0.32	0.155	0.165	0.23	0.155	0.25	0.29	0.155	0.23	0.27	0.140	0.140	0.195	0.140	0.22	0.26	0.135	0.29	0.32
600	0.120	0.150	0.27	0.31	0.130	0.160	0.21	0.125	0.25	0.28	0.125	0.22	0.26	0.110	0.140	0.180	0.110	0.22	0.24	0.110	0.29	0.31
740	0.099	–	–	–	0.105	0.160	0.190	0.100	0.25	0.27	–	–	–	0.094	0.135	0.165	0.094	0.21	0.23	0.089	0.29	0.30
960	0.075	–	–	–	0.086	0.155	0.180	0.082	0.24	0.26	–	–	–	0.077	0.135	0.155	0.077	0.21	0.22	0.071	0.29	0.29
1200	0.060	–	–	–	0.074	0.155	0.170	0.068	0.24	0.25	–	–	–	0.066	0.135	0.150	0.066	0.21	0.22	0.059	0.28	0.29

NOTE: Spacings larger than those specified in Method 12 (see table 4A) will result in larger voltage drop.

TABLE 4K2A

Multicore p.v.c.-insulated cables, non-armoured (ALUMINIUM CONDUCTORS)

BS 6004
BS 6346

Ambient temperature: 30°C
Conductor operating temperature: 70°C

CURRENT-CARRYING CAPACITY (Amperes):

Conductor cross-sectional area	Reference Method 4 (enclosed in an insulating wall, etc.)		Reference Method 3 (enclosed in conduit on a wall or ceiling, or in trunking)		Reference Method 1 (clipped direct)		Reference Method 11 (on a perforated cable tray), or Reference Method 13 (free air)	
	1 two-core cable, single-phase a.c. or d.c.	1 three-or four-core cable, three-phase a.c.	1 two-core cable, single-phase a.c. or d.c.	1 three- or four-core cable, three-phase a.c.	1 two-core cable, single-phase a.c. or d.c.	1 three- or four-core cable, three-phase a.c.	1 two-core cable, single-phase a.c. or d.c.	1 three- or four-core cable, three-phase a.c.
1	2	3	4	5	6	7	8	9
mm²	A	A	A	A	A	A	A	A
16	44	41	54	48	66	59	73	61
25	58	53	71	62	83	73	89	78
35	71	65	86	77	103	90	111	96
50	86	78	104	92	125	110	135	117
70	108	98	131	116	160	140	173	150
95	130	118	157	139	195	170	210	183
120	—	135	—	160	—	197	—	212
150	—	155	—	184	—	227	—	245
185	—	176	—	210	—	259	—	280
240	—	207	—	248	—	305	—	330
300	—	237	—	285	—	351	—	381

NOTE: Where the conductor is to be protected by a semi-enclosed fuse to BS 3036, see item 6.2 of the preface to this appendix.

TABLE 4K2B

VOLTAGE DROP (per ampere per metre): Conductor operating temperature: 70°C

Conductor cross-sectional area	Two-core cable d.c.	Two-core cable single-phase a.c.			Three- or four-core cable three-phase a.c.		
1	2	3			4		
mm²	mV	mV			mV		
16	4.5	4.5			3.9		
		r	x	z	r	x	z
25	2.9	2.9	0.175	2.9	2.5	0.150	2.5
35	2.1	2.1	0.170	2.1	1.80	0.150	1.80
50	1.55	1.55	0.170	1.55	1.35	0.145	1.35
70	1.05	1.05	0.165	1.05	0.90	0.140	0.92
95	0.77	0.77	0.160	0.79	0.67	0.140	0.68
120	—	—	—	—	0.53	0.135	0.55
150	—	—	—	—	0.42	0.135	0.44
185	—	—	—	—	0.34	0.135	0.37
240	—	—	—	—	0.26	0.130	0.30
300	—	—	—	—	0.21	0.130	0.25

TABLE 4K3A

Single-core armoured p.v.c.-insulated cables (non-magnetic armour)
(ALUMINIUM CONDUCTORS)

BS 6346

Ambient temperature: 30°C
Conductor operating temperature: 70°C

CURRENT-CARRYING CAPACITY (Amperes):

Conductor cross-sectional area	Reference Method 1 (clipped direct)		Reference Method 11 (on a perforated cable tray)		Reference Method 12 (free air)					
	2 cables single-phase a.c. or d.c. flat and touching	3 or 4 cables three-phase a.c. flat and touching	2 cables single-phase a.c. or d.c. flat and touching	3 or 4 cables three-phase a.c. flat and touching	2 cables single-phase a.c.		2 cables d.c. spaced		3 or 4 cables, three-phase a.c.		
					Horizontal flat spaced	Vertical flat spaced	Horizontal	Vertical	Horizontal flat spaced	Vertical flat spaced	3 cables trefoil
1	2	3	4	5	6	7	8	9	10	11	12
mm²	A	A	A	A	A	A	A	A	A	A	A
50	143	133	152	141	168	159	167	157	169	155	131
70	183	168	194	178	212	200	214	202	213	196	168
95	221	202	234	214	259	245	261	247	255	236	205
120	255	233	270	246	299	285	303	288	293	272	238
150	294	267	310	282	340	323	349	333	335	312	275
185	334	303	352	319	389	371	400	382	379	354	315
240	393	354	413	374	457	437	472	452	443	415	372
300	452	405	474	427	520	498	545	523	505	475	430
380	518	452	543	479	583	559	638	613	551	518	497
480	586	501	616	534	655	629	742	715	604	568	568
600	658	550	692	589	724	696	859	828	656	618	642
740	728	596	769	642	802	770	986	952	707	666	715
960	819	651	868	706	866	832	1171	1133	770	726	808
1200	893	692	952	756	938	902	1360	1317	822	774	880

NOTE: Where the conductor is to be protected by a semi-enclosed fuse to BS 3036, see item 6.2 of the preface to this appendix.

TABLE 4K3B

VOLTAGE DROP (per ampere per metre):   Conductor operating temperature: 70°C

Conductor cross-sectional area	2 cables d.c.	2 cables — single-phase a.c.						3 or 4 cables — three-phase a.c.								
		Reference Methods 1 & 11 (Touching)			Reference Method 12 (Spaced*)			Reference Methods 1, 11 and 12 (In trefoil touching)			Reference Methods 1 & 11 (Flat touching)			Reference Method 12 (Flat spaced*)		
		3			4			5			6			7		
mm²	mV	mV			mV			mV			mV			mV		
1	2	r	x	z	r	x	z	r	x	z	r	x	z	r	x	z
50	1.55	1.55	0.23	1.55	1.55	0.31	1.55	1.35	0.195	1.35	1.35	0.27	1.35	1.30	0.34	1.35
70	1.05	1.05	0.22	1.10	1.05	0.30	1.10	0.92	0.190	0.93	0.93	0.26	0.96	0.95	0.33	1.00
95	0.77	0.78	0.21	0.81	0.81	0.29	0.86	0.68	0.185	0.70	0.70	0.25	0.75	0.73	0.32	0.80
120	0.61	0.62	0.21	0.66	0.65	0.29	0.71	0.54	0.180	0.57	0.57	0.25	0.62	0.60	0.32	0.68
150	0.49	0.50	0.20	0.54	0.53	0.28	0.60	0.44	0.175	0.47	0.46	0.24	0.52	0.50	0.31	0.58
185	0.39	0.41	0.195	0.45	0.44	0.28	0.52	0.35	0.170	0.39	0.38	0.24	0.45	0.42	0.30	0.51
240	0.30	0.32	0.190	0.37	0.34	0.27	0.44	0.28	0.165	0.32	0.30	0.23	0.38	0.33	0.29	0.44
300	0.24	0.26	0.185	0.32	0.28	0.26	0.39	0.22	0.160	0.27	0.24	0.23	0.34	0.28	0.29	0.40
380	0.190	0.22	0.185	0.28	0.26	0.25	0.36	0.185	0.155	0.24	0.22	0.22	0.32	0.27	0.26	0.38
480	0.150	0.180	0.180	0.25	0.22	0.25	0.33	0.155	0.155	0.22	0.195	0.22	0.29	0.24	0.25	0.35
600	0.120	0.150	0.175	0.23	0.195	0.24	0.31	0.130	0.150	0.200	0.170	0.21	0.27	0.21	0.24	0.32
740	0.097	0.135	0.170	0.22	0.180	0.23	0.29	0.115	0.145	0.185	0.160	0.20	0.26	0.200	0.22	0.30
960	0.075	0.115	0.160	0.200	0.165	0.21	0.27	0.100	0.140	0.175	0.150	0.185	0.24	0.190	0.195	0.27
1200	0.060	0.110	0.155	0.190	0.160	0.180	0.24	0.094	0.140	0.170	0.145	0.160	0.22	0.185	0.165	0.25

*NOTE: Spacings larger than those specified in Method 12 (see table 4A) will result in larger voltage drop.

TABLE 4K4A

Multicore armoured p.v.c.-insulated cables (ALUMINIUM CONDUCTORS)

BS 6346

Ambient temperature: 30°C
Conductor operating temperature: 70°C

CURRENT-CARRYING CAPACITY (Amperes):

Conductor cross-sectional area	Reference Method 1 (clipped direct)		Reference Method 11 (on a perforated cable tray) or Reference Method 13 (free air)	
	1 two-core cable, single-phase a.c. or d.c.	1 three- or four-core cable, three-phase a.c.	1 two-core cable, single-phase a.c. or d.c.	1 three- or four-core cable, three-phase a.c.
1	2	3	4	5
mm²	A	A	A	A
16	68	58	71	61
25	89	76	94	80
35	109	94	115	99
50	131	113	139	119
70	165	143	175	151
95	199	174	211	186
120	—	202	—	216
150	—	232	—	250
185	—	265	—	287
240	—	312	—	342
300	—	360	—	399

NOTE: Where the conductor is to be protected by a semi-enclosed fuse to BS 3036, see item 6.2 of the preface to this appendix.

TABLE 4K4B

VOLTAGE DROP (per ampere per metre): Conductor operating temperature: 70°C

Conductor cross-sectional area	Two-core cable d.c.	Two-core cable single-phase a.c.			Three- or four-core cable three-phase a.c.		
1	2	3			4		
mm²	mV	mV			mV		
16	4.5	4.5			3.9		
		r	x	z	r	x	z
25	2.9	2.9	0.175	2.9	2.5	0.150	2.5
35	2.1	2.1	0.170	2.1	1.80	0.150	1.80
50	1.55	1.55	0.170	1.55	1.35	0.145	1.35
70	1.05	1.05	0.165	1.05	0.90	0.140	0.92
95	0.77	0.77	0.160	0.79	0.67	0.140	0.68
120	—	—	—	—	0.53	0.135	0.55
150	—	—	—	—	0.42	0.135	0.44
185	—	—	—	—	0.34	0.135	0.37
240	—	—	—	—	0.26	0.130	0.30
300	—	—	—	—	0.21	0.130	0.25

TABLE 4L1A

Single-core thermosetting insulated cables, non-armoured (ALUMINIUM CONDUCTORS)

BS 5467

Ambient temperature: 30°C
Conductor operating temperature: 90°C

CURRENT-CARRYING CAPACITY (Amperes):

Conductor cross-sectional area	Reference Method 4 (enclosed in conduit in thermally insulating wall etc.)		Reference Method 3 (enclosed in conduit on a wall or in trunking etc.)		Reference Method 1 (clipped direct)		Reference Method 11, (on a perforated cable tray horizontal or vertical)		Reference Method 12 (free air)		
	2 cables single-phase a.c. or d.c.	3 or 4 cables three-phase a.c.	2 cables, single-phase a.c. or d.c.	3 or 4 cables three-phase a.c.	2 cables, single-phase a.c. or d.c. flat & touching	3 or 4 cables three-phase a.c. flat & touching or trefoil	2 cables, single-phase a.c. or d.c. flat & touching	3 or 4 cables, three-phase a.c. flat & touching or trefoil	Horizontal flat spaced 2 cables, single-phase a.c. or d.c. or 3 three-phase a.c.	Vertical flat spaced 2 cables, single-phase a.c. or d.c. or 3 three-phase a.c.	Trefoil 3 cables, trefoil three-phase a.c.
1	2	3	4	5	6	7	8	9	10	11	12
mm²	A	A	A	A	A	A	A	A	A	A	A
50	125	113	157	140	169	149	180	165	210	188	159
70	158	142	200	179	215	189	231	211	271	244	206
95	191	171	242	217	265	234	281	258	332	300	253
120	220	197	281	251	308	273	326	300	387	351	296
150	253	226	—	—	353	314	376	346	448	408	343
185	288	256	—	—	410	366	430	396	515	470	395
240	338	300	—	—	489	438	509	469	611	561	471
300	387	344	—	—	564	507	586	541	708	652	544
380	—	—	—	—	658	594	679	628	798	742	638
480	—	—	—	—	765	692	786	728	927	865	743
600	—	—	—	—	871	791	903	836	1058	990	849
740	—	—	—	—	1001	911	1025	951	1218	1143	979
960	—	—	—	—	1176	1072	1191	1108	1440	1355	1151
1200	—	—	—	—	1333	1217	1341	1249	1643	1550	1307

NOTE: 1. Where the conductor is to be protected by a semi-enclosed fuse to BS 3036, see item 6.2 of the preface to this appendix.
2. Where a conductor operates at a temperature exceeding 70°C it shall be ascertained that the equipment connected to the conductor is suitable for the conductor operating temperature (see Regulation 512-02).

TABLE 4L1B

VOLTAGE DROP (per ampere per metre): Conductor operating temperature: 90°C

Conductor cross-sectional area	2 cables d.c.	2 cables – single-phase a.c.											3 or 4 cables – three-phase a.c.										
		Reference Methods 3 & 4 (Enclosed in conduit etc., in or on a wall)			Reference Methods 1 & 11 (Clipped direct or on trays, touching)			Reference Method 12 (Spaced*)			Reference Methods 3 & 4 (Enclosed in conduit etc., or on a wall)			Reference Methods 1, 11 & 12 (In trefoil)			Reference Methods 1 & 11 (Flat and touching)			Reference Method 12 (Flat spaced*)		
1	2	3			4			5			6			7			8			9		
mm²	mV	mV			mV			mV			mV			mV			mV			mV		
		r	x	z	r	x	z	r	x	z	r	x	z	r	x	z	r	x	z	r	x	z
50	1.65	1.70	0.30	1.72	1.65	0.190	1.66	1.65	0.28	1.68	1.44	0.26	1.46	1.44	0.165	1.45	1.44	0.24	1.46	1.44	0.32	1.48
70	1.13	1.17	0.30	1.21	1.12	0.185	1.14	1.12	0.27	1.15	1.00	0.26	1.04	0.97	0.160	0.98	0.97	0.24	1.00	0.97	0.31	1.02
95	0.82	0.86	0.29	0.91	0.82	0.185	0.84	0.82	0.27	0.94	0.75	0.25	0.79	0.71	0.160	0.73	0.71	0.23	0.75	0.71	0.31	0.78
120	0.65	0.68	0.29	0.74	0.65	0.180	0.67	0.65	0.27	0.70	0.59	0.25	0.64	0.57	0.155	0.59	0.57	0.23	0.61	0.57	0.31	0.64
150	0.53	0.54	0.28	0.61	0.52	0.175	0.55	0.52	0.26	0.58	0.48	0.24	0.54	0.45	0.155	0.47	0.45	0.23	0.50	0.45	0.30	0.54
185	0.42	0.45	0.28	0.53	0.43	0.175	0.46	0.42	0.26	0.49	0.38	0.24	0.45	0.36	0.150	0.39	0.36	0.23	0.43	0.36	0.30	0.47
240	0.32	0.34	0.27	0.43	0.32	0.170	0.36	0.32	0.26	0.41	0.30	0.24	0.38	0.28	0.150	0.32	0.28	0.22	0.35	0.28	0.30	0.41
300	0.26	0.28	0.27	0.38	0.26	0.170	0.31	0.26	0.26	0.36	0.25	0.23	0.34	0.22	0.145	0.27	0.22	0.22	0.31	0.22	0.30	0.37
380	0.20	—	—	—	0.21	0.165	0.27	0.21	0.25	0.33	0.20	0.23	0.31	0.180	0.145	0.23	0.180	0.22	0.28	0.180	0.29	0.34
480	0.160	—	—	—	0.170	0.165	0.23	0.165	0.25	0.30	0.165	0.23	0.28	0.150	0.140	0.20	0.150	0.22	0.27	0.145	0.29	0.32
600	0.130	—	—	—	0.140	0.160	0.21	0.135	0.25	0.28	0.135	0.22	0.26	0.120	0.140	0.185	0.120	0.22	0.25	0.120	0.29	0.31
740	0.105	—	—	—	0.115	0.160	0.19	0.110	0.25	0.27	—	—	—	0.100	0.135	0.170	0.100	0.21	0.23	0.095	0.29	0.30
960	0.080	—	—	—	0.092	0.155	0.18	0.087	0.24	0.26	—	—	—	0.082	0.135	0.160	0.082	0.21	0.23	0.076	0.29	0.30
1200	0.064	—	—	—	0.079	0.155	0.17	0.073	0.24	0.25	—	—	—	0.070	0.135	0.150	0.070	0.21	0.22	0.063	0.28	0.29

*NOTE: Spacings larger than those specified in Method 12 (see table 4A) will result in larger voltage drop.

TABLE 4L2A

**Multicore thermosetting insulated cables non-armoured
(ALUMINIUM CONDUCTORS)**

BS 5467

Ambient temperature: 30°C
Conductor operating temperature: 90°C

CURRENT-CARRYING CAPACITY (Amperes):

Conductor cross-sectional area	Reference Method 4 (enclosed in an insulated wall, etc.)		Reference Method 3 (enclosed in conduit on a wall or ceiling, or in trunking)		Reference Method 1 (clipped direct)		Reference Method 11 (on a perforated cable tray), or Reference Method 13 (free air)	
	1 two-core cable, single-phase a.c. or d.c.	1 three- or four-core cable, three-phase a.c.	1 two-core cable, single-phase a.c. or d.c.	1 three- or four-core cable, three-phase a.c.	1 two-core cable, single-phase a.c. or d.c.	1 three- or four-core cable, three-phase a.c.	1 two-core cable, single-phase a.c. or d.c.	1 three- or four-core cable, three-phase a.c.
1	2	3	4	5	6	7	8	9
mm²	A	A	A	A	A	A	A	A
16	60	55	72	64	84	76	91	77
25	78	71	94	84	101	90	108	97
35	96	87	115	103	126	112	135	120
50	115	104	138	124	154	136	164	146
70	145	131	175	156	198	174	211	187
95	175	157	210	188	241	211	257	227
120	—	180	—	216	—	245	—	263
150	—	206	—	240	—	283	—	304
185	—	233	—	272	—	323	—	347
240	—	273	—	318	—	382	—	409
300	—	313	—	364	—	440	—	471

NOTE: 1. Where the conductor is to be protected by a semi-enclosed fuse to BS 3036, see item 6.2 of the preface to this appendix.
2. Where a conductor operates at a temperature exceeding 70°C it shall be ascertained that the equipment connected to the conductor is suitable for the conductor operating temperature (see Regulation 512-02).

TABLE 4L2B

VOLTAGE DROP (per ampere per metre):   Conductor operating temperature: 90°C

Conductor cross-sectional area	Two-core cable d.c.	Two-core cable single-phase a.c.			Three or four-core cable three-phase a.c.		
1	2	3			4		
mm²	mV	mV			mV		
16	4.8	4.8			4.2		
		r	x	z	r	x	z
25	3.1	3.1	0.165	3.1	2.7	0.140	2.7
35	2.2	2.2	0.160	2.2	1.90	0.140	1.95
50	1.60	1.65	0.160	1.65	1.40	0.135	1.45
70	1.10	1.10	0.155	1.15	0.96	0.135	0.97
95	0.82	0.82	0.150	0.84	0.71	0.130	0.72
120	—	—	—	—	0.56	0.130	0.58
150	—	—	—	—	0.45	0.130	0.47
185	—	—	—	—	0.37	0.130	0.39
240	—	—	—	—	0.28	0.125	0.31
300	—	—	—	—	0.23	0.125	0.26

TABLE 4L3A

Single-core cables having thermosetting insulation (non-magnetic armour)
(ALUMINIUM CONDUCTORS)

BS 5467
BS 6724

CURRENT-CARRYING CAPACITY (Amperes):

Ambient temperature: 30°C
Conductor operating temperature: 90°C

Conductor cross-sectional area	Reference Method 1 (clipped direct)		Reference Method 11 (on a perforated cable tray)		Reference Method 12 (free air)						
	2 cables, single-phase a.c. or d.c. flat and touching	3 or 4 cables, three-phase a.c. flat and touching	2 cables, single-phase a.c. or d.c. flat and touching	3 or 4 cables, three-phase a.c. flat and touching	2 cables single-phase a.c.		2 cables d.c.		3 or 4 cables, three-phase a.c.		
					Horizontal flat spaced	Vertical flat spaced	Horizontal spaced	Vertical spaced	Horizontal flat spaced	Vertical flat spaced	3 cables trefoil
1	2	3	4	5	6	7	8	9	10	11	12
mm²	A	A	A	A	A	A	A	A	A	A	A
50	179	165	192	176	212	199	216	197	215	192	162
70	228	209	244	222	269	254	275	253	270	244	207
95	276	252	294	267	328	310	332	307	324	296	252
120	320	291	340	308	378	358	384	357	372	343	292
150	368	333	390	352	429	409	441	411	424	394	337
185	419	378	444	400	490	467	511	480	477	447	391
240	494	443	521	468	576	549	605	572	554	523	465
300	568	508	597	536	654	624	701	666	626	595	540
380	655	573	688	608	735	704	812	780	639	649	625
480	747	642	786	685	825	790	942	906	765	717	714
600	836	706	880	757	909	872	1076	1036	832	780	801
740	934	764	988	824	989	950	1250	1205	890	835	897
960	1056	838	1121	911	1094	1052	1488	1435	970	911	1014
1200	1163	903	1236	990	1187	1141	1715	1658	1043	980	1118

NOTE: 1. Where the conductor is to be protected by a semi-enclosed fuse to BS 3036, see item 6.2 of the preface to this appendix.
2. Where a conductor operates at a temperature exceeding 70°C it shall be ascertained that the equipment connected to the conductor is suitable for the conductor operating temperature (see Regulation 512-02).

TABLE 4L3B

VOLTAGE DROP (per ampere per metre): Conductor operating temperature: 90°C

Conductor cross-sectional area	2 cables d.c.	2 cables — single-phase a.c.							3 or 4 cables — three-phase a.c.						
		Reference Methods 1 & 11 (Touching)			Reference Method 12 (Spaced*)			Reference Methods 1, 11 and 12 (In trefoil touching)			Reference Methods 1 & 11 (Flat touching)			Reference Method 12 (Flat spaced*)		
1	2	3			4			5			6			7		
mm²	mV	mV			mV			mV			mV			mV		
		r	x	z	r	x	z	r	x	z	r	x	z	r	x	z
50	1.60	1.60	0.22	1.60	1.60	0.30	1.60	1.40	0.185	1.40	1.40	0.26	1.40	1.35	0.34	1.40
70	1.10	1.10	0.21	1.15	1.10	0.29	1.15	0.96	0.180	0.98	0.97	0.25	1.00	0.99	0.33	1.05
95	0.82	0.83	0.20	0.85	0.85	0.29	0.90	0.71	0.175	0.74	0.74	0.25	0.78	0.76	0.32	0.83
120	0.66	0.66	0.20	0.69	0.69	0.28	0.74	0.57	0.170	0.60	0.60	0.24	0.64	0.63	0.31	0.70
150	0.52	0.53	0.195	0.57	0.56	0.28	0.62	0.46	0.170	0.49	0.49	0.24	0.54	0.52	0.30	0.60
185	0.42	0.43	0.190	0.47	0.46	0.27	0.54	0.38	0.165	0.41	0.40	0.24	0.47	0.44	0.30	0.53
240	0.32	0.34	0.185	0.39	0.37	0.27	0.45	0.29	0.160	0.34	0.32	0.23	0.39	0.35	0.29	0.46
300	0.26	0.27	0.185	0.33	0.30	0.26	0.40	0.24	0.160	0.29	0.26	0.23	0.34	0.29	0.29	0.41
380	0.21	0.23	0.180	0.29	0.26	0.25	0.36	0.195	0.155	0.25	0.23	0.22	0.32	0.27	0.27	0.38
480	0.160	0.185	0.175	0.25	0.23	0.25	0.34	0.160	0.155	0.22	0.20	0.21	0.29	0.24	0.26	0.35
600	0.130	0.160	0.175	0.24	0.20	0.24	0.31	0.135	0.150	0.20	0.175	0.21	0.27	0.22	0.25	0.33
740	0.105	0.140	0.170	0.22	0.190	0.22	0.29	0.120	0.145	0.190	0.165	0.195	0.26	0.21	0.22	0.30
960	0.080	0.120	0.160	0.20	0.170	0.21	0.27	0.105	0.140	0.175	0.150	0.180	0.24	0.195	0.195	0.28
1200	0.064	0.105	0.160	0.190	0.155	0.20	0.25	0.093	0.135	0.165	0.140	0.175	0.22	0.180	0.185	0.26

*NOTE: Spacings larger than those specified in Method 12 (see table 4A) will result in larger voltage drop.

TABLE 4L4A

**Multicore armoured cables having thermosetting insulation
(ALUMINIUM CONDUCTORS)**

BS 5467
BS 6724

Ambient temperature: 30°C
Conductor operating temperature: 90°C

CURRENT-CARRYING CAPACITY (Amperes):

Conductor cross-sectional area	Reference Method 1 (clipped direct)	Reference Method 1 (clipped direct)	Reference Method 11 (on a perforated cable tray) or Reference Method 13 (free air)	Reference Method 11 (on a perforated cable tray) or Reference Method 13 (free air)
	1 two-core cable, single-phase a.c. or d.c.	1 three- or four-core cable, three-phase a.c.	1 two-core cable, single-phase a.c. or d.c.	1 three- or four-core cable three-phase a.c.
1	2	3	4	5
mm^2	A	A	A	A
16	82	71	85	74
25	108	92	112	98
35	132	113	138	120
50	159	137	166	145
70	201	174	211	185
95	242	214	254	224
120	—	249	—	264
150	—	284	—	305
185	—	328	—	350
240	—	386	—	418
300	—	441	—	488

NOTES: 1. Where the conductor is to be protected by a semi-enclosed fuse to BS 3036, see item 6.2 of the preface to this appendix.
2. Where a conductor operates at a temperature exceeding 70°C it shall be ascertained that the equipment connected to the conductor is suitable for the conductor operating temperature (see Regulation 512-02).

TABLE 4L4B

VOLTAGE DROP (per ampere per metre): Conductor operating temperature: 90°C

Conductor cross-sectional area	Two-core cable d.c.		Two-core cable single-phase a.c.			Three or four-core cable three-phase a.c.		
1	2		3			4		
mm²	mV		mV			mV		
16	4.8		4.8			4.2		
		r	r	x	z	r	x	z
25	3.1	3.1	3.1	0.165	3.1	2.7	0.140	2.7
35	2.2	2.2	2.2	0.160	2.2	1.90	0.140	1.95
50	1.60	1.65	1.65	0.160	1.65	1.40	0.135	1.45
70	1.10	1.10	1.10	0.155	1.15	0.96	0.135	0.97
95	0.82	0.82	0.82	0.150	0.84	0.71	0.130	0.72
120	—		—			0.56	0.130	0.58
150	—		—			0.45	0.130	0.47
185	—		—			0.37	0.130	0.39
240	—		—			0.28	0.125	0.31
300	—		—			0.23	0.125	0.26

APPENDIX 5

CLASSIFICATION OF EXTERNAL INFLUENCES
(See note to Chapter 32)

This appendix gives the classification and codification of external influences developed for IEC Publication 364.

Each condition of external influence is designated by a code comprising a group of two capital letters and a number, as follows:

The first letter relates to the general category of external influence:

A.   Environment

B.   Utilisation

C.   Construction of buildings

The second letter relates to the nature of the external influence:

   ... *A.*

   ... *B.*

   ... *C.*

The number relates to the class within each external influence:

     1

     2

     3

For example the code AA4 signifies:

A   = Environment

AA   = Environment − Ambient temperature

AA4 = Environment − Ambient temperature in the range of −5°C to +40°C.

> NOTE − The codification given in this appendix is not intended to be used for marking equipment.

APPENDIX 5

CONCISE LIST OF EXTERNAL INFLUENCES

Environment	A	**AA**	*Ambient (°C)*	**AF**	*Corrosion*	**AM**	*Radiation*	
		AA1	−60°C +5°C	AF1	Negligible	AM1	Negligible	
		AA2	−40°C +5°C	AF2	Atmospheric	AM2	Stray currents	
		AA3	−25°C +5°C	AF3	Intermittent	AM3	Electromagnetic	
		AA4	−5°C +40°C	AF4	Continuous	AM4	Ionization	
		AA5	+5°C +40°C			AM5	Electrostatics	
		AA6	+5°C +60°C	**AG**	*Impact*	AM6	Induction	
		AB	*Humidity*	AG1	Low	**AN**	*Solar*	
				AG2	Medium			
		AC	*Altitude (metres)*	AG3	High	AN1	Negligible	
						AN2	Significant	
		AC1	≤ 2,000 metres	**AH**	*Vibration*			
		AC2	> 2,000 metres					
				AH1	Low	**AP**	*Seismic*	
		AD	*Water*	AH2	Medium	AP1	Negligible	
				AH3	High	AP2	Low	
		AD1	Negligible			AP3	Medium	
		AD2	Drops	**AJ**	*Other mechanical stresses*	AP4	High	
		AD3	Sprays					
		AD4	Splashes	**AK**	*Flora*	**AQ**	*Lightning*	
		AD5	Jets					
		AD6	Waves	AK1	No hazard	AQ1	Negligible	
		AD7	Immersion	AK2	Hazard	AQ2	Indirect	
		AD8	Submersion			AQ3	Direct	
				AL	*Fauna*			
		AE	*Foreign bodies*			**AR**	*Wind*	
				AL1	No hazard			
		AE1	Negligible	AL2	Hazard			
		AE2	Small					
		AE3	Very small					
		AE4	Dust					
Utilization	B	**BA**	*Capability*	**BD**	*Evacuation*	**BE**	*Materials*	
		BA1	Ordinary	BD1	(Low density/ easy exit)	BE1	No risk	
		BA2	Children			BE2	Fire risk	
		BA3	Handicapped	BD2	(Low density/ difficult exit)	BE3	Explosion risk	
		BA4	Instructed			BE4	Contamination risk	
		BA5	Skilled	BD3	(High density/ easy exit)			
		BB	*Resistance*	BD4	(High density/ difficult exit)			
		BC	*Contact with earth*					
		BC1	None					
		BC2	Low					
		BC3	Frequent					
		BC4	Continuous					
Building	C	**CA**	*Materials*	**CB**	*Structure*			
		CA1	Non-combustible	CB1	Negligible risk			
		CA2	Combustible	CB2	Fire propagation			
				CB3	Structure movement			
				CB4	Flexible			

APPENDIX 6

FORMS OF COMPLETION AND INSPECTION CERTIFICATE

Introduction

i) The forms of Completion, Inspection and Testing required by Part 7 shall be made out and signed by competent persons in respect of the design, construction, inspection and testing of the work.

ii) Competent persons will, as appropriate to their function under i) above, have a sound knowledge and experience relevant to the nature of the installation undertaken and to the technical standards set down in the Wiring Regulations, be fully versed in the inspection and testing procedures contained in the Regulations and employ adequate testing equipment.

iii) Completed forms will indicate the responsibility for design, construction, inspection and testing, whether in relation to new work or further work on an existing installation.

iv) When making out and signing a form on behalf of a company or other business entity, individuals shall state for whom they are acting.

v) Additional forms may be required as clarification, if needed by non-technical persons, or in expansion, for larger or more complicated jobs.

FORMS OF COMPLETION AND INSPECTION CERTIFICATE
(as prescribed in the IEE Regulations for Electrical Installations)

(1.) (see Notes overleaf)

DETAILS OF THE INSTALLATION

Client:

Address:

DESIGN

I/We being the person(s) responsible (as indicated by my/our signatures below) for the Design of the electrical installation, particulars of which are described on Page 3 of this form CERTIFY that the said work for which I/we have been responsible is to the best of my/our knowledge and belief in accordance with the Regulations for Electrical Installations published by the Institution of Electrical Engineers, 16th Edition, amended to (3.) (date ....) except for the departures, if any, stated in this Certificate.

The extent of liability of the signatory is limited to the work described above as the subject of this Certificate.

For the DESIGN of the installation:

Name (In Block Letters):   Position:

For and on behalf of:

Address:

(2.) Signature: (3.) Date

CONSTRUCTION

I/We being the person(s) responsible (as indicated by my/our signatures below) for the Construction of the electrical installation, particulars of which are described on Page 3 of this form CERTIFY that the said work for which I/we have been responsible is to the best of my/our knowledge and belief in accordance with the Regulations for Electrical Installations published by the Institution of Electrical Engineers, 16th Edition, amended to (3.) (date ....) except for the departures, if any, stated in this Certificate.

The extent of liability of the signatory is limited to the work described above as the subject of this Certificate.

For the CONSTRUCTION of the installation:

Name (In Block Letters):   Position:

For and on behalf of:

Address:

(2.) Signature: (3.) Date:

INSPECTION AND TEST

I/We being the person(s) responsible (as indicated by my/our signatures below) for the Inspection and Test of the electrical installation, particulars of which are described on Page 3 of this form CERTIFY that the said work for which I/we have been responsible is to the best of my/our knowledge and belief in accordance with the Regulations for Electrical Installations published by the Institution of Electrical Engineers, 16th Edition, amended on (3.) (date ....) except for departures, if any, stated in this Certificate.

The extent of liability of the signatory is limited to the work described above as the subject of this Certificate.

For the INSPECTION AND TEST of the installation:

Name (In Block Letters):   Position:

For and on behalf of:

Address:

I RECOMMEND that this installation be further inspected and tested after an interval of not more than .... years. (5.)

(2.) Signature: (3.) Date:

(6) page 1 of .... pages

1. This document is intended for the initial certification of a new installation or of an alteration or addition to an existing installation and of an inspection.

2. The signatures appended are those of the persons authorised by the companies executing the work of design, construction and inspection and testing respectively. A signatory authorised to certify more than one category of work shall sign in each of the appropriate places.

3. Dates to be inserted.

4. Where particulars of the installation recorded herein constitute a sufficient schedule for the purpose of Regulation 514-09-01 further drawings/schedules need not be provided. For other installations the additional drawings/schedules listed below apply.

5. Insert here the time interval recommended between periodic inspections. Regard should be paid to relevant National or Local legislation and reference should be made to Chapter 13.

6. The page numbers of each sheet should be indicated together with the total number of sheets involved.

(4) **Schedule of additional records.**

PARTICULARS OF THE INSTALLATION

(Delete or complete items as appropriate)

Type of Installation   New/alteration/addition/to existing installation

Type of Earthing (312-03):   TN-C   TN-S   TN-C-S   TT   IT
(Indicate in the box)   ☐   ☐   ☐   ☐   ☐

Earth Electrode: Resistance .... ohms

                       Method of Measurement ....

                       Type (542-02-01) and Location ....

Characteristics of the supply at the origin of the installation (313-01):

         Nominal voltage .... volts

         Frequency .... Hz   Number of phases ....

ascertained by enquiry	determined by calculation	measured

         Prospective short-circuit current .... kA

         Earth fault loop impedance (Z_e) ....ohms

         Maximum demand ....A per phase

         Overcurrent protective device - Type BS .... Rating ..... A

Main switch or circuit-breaker (460-01-02): Type BS .... Rating ..... A No of poles .....

         (if an r.c.d., rated residual operating current $I_{\Delta n}$ ....mA.)

Method of protection against indirect contact:

1. Earthed equipotential bonding and automatic disconnection of supply ☐

or

2. Other ☐ (Describe) ....

Main equipotential bonding conductors (413-02-01/02, 547-02-01): Size.......... ..mm²

Schedule of Test Results: Continuation .... pages

Details of departures (if any) from the Wiring Regulations (120-04, 120-05).................

Comments on existing installation, where applicable (743-01-01): ....

(6) page 3 of .... pages

Index

A

Abrasion of cables	526-06, 522-08
A.C. types of cables suitable for	521-02-01
Accessibility —	
— connections	526-04
— emergency switching	537-04
— equipment	130-07, 513
— protective conductor connections	543-03-03
Accessories —	
— concealed cables to	522-06
— selection and erection of	Sec. 553
Accessory, definition	p. 7
Additions to installations	130-09, Chap. 72
Adverse conditions, precautions in	Sec. 522
Aerial cables	521-01
Agricultural and horticultural installations —	p. 110
— applicability of the Regulations to	110-01
— electric fences	605-14
— protective measures against shock	
— statutory regulations	Appx. 2
— wiring systems in livestock locations	522-10
Air conditioning systems, main bonding of	413-02
Aircraft, electrical equipment of, the Regulations not applicable to	110-02
Alarm —	
— devices, safety services	563-01
— fire (*See Fire alarm*)	
Alterations to installations	130-09, Chap. 72
Aluminium —	
— conductors —	
— as bonding connections	547-01
— in contact with brass etc.	522-05
— prohibited for connections to earth electrodes	542-03
Ambient temperature —	
— plugs and socket-outlets	553-01
— cables and conductors	522-01, Appx. 4
— cables —	
— correction factors	Appx. 4
— determination of	Appx. 4
— definition	p. 7
Amendments to the Regulations	Preface, p. vi
Appendix	p. 151
Appliance, definition	p. 7
Appliances —	p. 9
— fire hazard from	422
— flexible cables for	521-01
— heating —	
— bathrooms	Sec. 601
— portable (*See also Portable equipment*)	
— portable, connection of	553-01
— stationary, in bathrooms	Sec. 601
— switches mounted on	476-03
— switching for	476-03
Application of protective measures for safety	Chap. 47
Arcades, discharge lighting in	476-03
Arcing of switchgear	130-03, 422-01
Arm's reach, definition	p. 7
Arm's reach,	412-05
Automatic disconnection of supply, protection by (*See Earthed equipotential bonding and automatic disconnection*)	
Automatic sources, safety services supplies	313-02, Sec. 562
Autotransformers, use of	551-01
Auxiliary supply for residual current device	531-02

B

Bare conductors —	
— as overhead lines	412-05
— cables connected to	523-03
— current-carrying capacity	523-03
— electromechanical stresses	521-03
— identification	514-06
— mechanical damage	522-06, 522-08
— switchboards, on	523-03
— voltage drop	525-01
Bare live parts in SELV circuits	411-02
Bare live parts, placing out of reach	412-05
Barriers —	
— fire	527-02
— protection by —	412-03
— application of	471-05, 471-05
— degrees of protection	412-03
— intermediate, in enclosures	412-03, 413-03
— openings in	412-03, 471-05
— provided during erection, testing	713-07
— removal or opening of	412-03
— securing of	412-03
— vertical wiring system runs	527-02
Basic insulation, definition	p. 7
Bathrooms, Protective measures for	Sec. 601
Baths, bonding of	Sec. 601
Batten lampholder, defined as luminaire	p.12
Batteries —	
— for safety services	Sec. 562
— Regulations not applicable to parts of	110-03
Bends —	
— cable ducts or ducting	522-08
— cables	522-08
— conduit systems	522-08
Body resistance —	
— conventional	471-01
— in relation to nominal voltage of SELV circuits	471-01
— in relation to protection by automatic disconnection	471-01
— livestock	Sec. 605
Bonding —	
— caravans	Sec. 608
— conductor, definition	p. 7
— conductors —	
— main	413-02, 547-02
— selection and erection of	Sec. 547
— supplementary	413-02, 547-03
— connections, warning notice at	514-13
— earth-free local	413-05, 471-11
— equipment in safety separation circuits	413-06

246

— 'instantaneous' water heaters	554-05
— local supplementary	413-02
— main, to other services and extraneous parts	413-02
— other services at points of contact	528-02
— other services, fundamental requirement	130-04
— single-core metal-sheathed cables	523-05
Boxes —	
— joints and terminations in	523-03 Sec 526
— non-metallic, for suspension of luminaires	Sec. 527
— segregation of circuits in	528-01
Breaking capacity —	
— fundamental requirement	130-03
— overload protective devices	432-02, 432-03
— short-circuit protective devices	432-02, 432-04, 434-03
British Standards —	
British Standards —	
— Codes of Practice	Preface p. viii, ix
— compliance with	Sec. 511
— Institution	Preface p. viii, Appx. 1
— reference in the Regulations, list	Appx. 1
British Telecom, earth wires not to be bonded	413-02
Building materials, fire hazard from equipment	422-01, 422-02
Building site (See Construction site)	
Bunched, definition	p. 8
Busbars and busbar connections—	
— cables connected to busbars	523-03
— colour identification of	Sec. 514
— current-carrying capacity	523-01
— selection of	521-01
— trunking systems, as protective conductors	543-02
Burns, protection against	Table 42A

C

Cable —	
— capacities of conduit and trunking	522-08
— channels, temperatures in	Sec. 422
— channels, waterproofing	522-03
— coupler, definition	p. 8
— couplers —	
— for joints in flexibles	526-04
— reduced low voltage circuits	471-15
— selection and connection of	Chap. 52
— covers	522-06
— ducting —	
— bends	522-08
— definition	p. 8
— fire barriers in	527-02
— joints in	526-01, 543-03
— protective conductors formed by	543-02
— selection of	521-05
— supports	522-08
— vertical, temperatures in	422-01
— duct, definition	p. 9

— ducts —	
— bends	522-08
— fire barriers in	527-02
— underground	522-06
— vertical, temperatures in	Sec. 422
— waterproofing	522-03
— enclosures —	
— a.c. circuits in steel	521-05
— cable capacities of	522-08
— trunking (See Trunking systems)	
Cables —	
— abrasion	522-06, 522-08
— a.c., types for	521-02
— aluminium-conductor —	
— as bonding connections	547-01
— prohibited for connections to earth electrodes	542-03
— ambient temperature	Appx. 4, 522-01
— armour as protective conductor	543-02
— attack by vermin	522-10
— bends	522-08
— buried direct in ground	522-06
— colour identification	Sec. 514
— concealed in walls or partitions	522-06
— concealed under floors or above ceilings	522-06
— connected to bare conductors or busbars	523-03
— core identification	Sec. 514
— corrosive or polluting substances in contact with	522-05
— current-carrying capacity	Appx. 4, 523-01 to 523-05
— damage by fauna	522-10
— electromagnetic effects	521-02
— electromechanical stresses	521-03
— emergency lighting, segregation	528-01
— enclosures, in onerous dust conditions	522-04
— environmental conditions	Sec. 522
— extra-low voltage	411-02
— fire-alarm, segregation	528-01
— fire barriers	527-02
— fixings (See Cables, Supports)	
— flexible (including flexible cords) —	
— appliances, for	521-01
— colour identification	514-07
— cords, applications of	Appx. 4
— current-carrying capacity	Appx. 4
— fixed equipment, connections to	526
— fixed wiring, use as	521-01
— identification of cores	514-07
— joints	526-01 & 526-02
— luminaires, supporting	Appx. 4
— portable equipment to be connected by	521-01
— safety separated circuits	413-06
— segregation of circuits in	528-01
— selection of types, LV	Appx. 4, 521-01
— floor-warming	554-06
— grouping	Appx. 4 (Table 4A)
— heating	554-06
— HV discharge lighting (See High Voltage)	
— identification or cores	Sec. 514
— joints, accessibility	526-04
— joints, general	Sec. 526

247

— lift shafts	528-02	— protective measures for	608-02 to 608-04, 608-10
— livestock situations	522-10	— warning notice in	608-07
— low temperatures, in	522-01	Cartridge fuse link,	
— mechanical stresses	Sec. 522	definition	p. 8
— metal coverings as protective conductors	543-02	Catenary wires	521-01
— metal-sheathed, single-core, bonding	523-05	Ceiling roses — — exemption from requirements for enclosures	412-03
— metal sheaths as earth electrodes	542-02	— multiple pendants	553-04
— methods of installation	Appx. 4 Table 4A	Ceilings, thermally insulated, cables above	523-04
— mineral-insulated — discharge lighting circuits	521-01	Ceilings, cables above, mechanical damage	522-06-05
— emergency lighting systems	528-01	Central heating risers, main bonding to	413-02
— fire alarm systems	528-01	Certificates —	
— sealing	522-03	— completion	741, 130-10, Appx. 6
— non-flexible, LV, selection of types	521-01	— forms of	Appx. 6
— non-metal-enclosed, as protection against direct and indirect contact	471-09	— inspection	732, Appx. 6
		— provision of	741, 742, 744
		Certification	Secs. 721, 732, 744
— non-sheathed, to be enclosed	522-03, 521-07	Characteristics, general, of installation	300-01
— numbers in enclosures	522-08	Charts (See Diagrams)	
— operational conditions	Sec. 522	Cinematograph installations	Appx. 2
— outdoors on walls	522-06	Circuit arrangements —	
— overhead between buildings etc.	412-05	— division of installation	314-01
— paralleled	433-03, 434-04, 523-02	Circuit-breaker, definition	p. 8
— reduced neutral	524-02	Circuit-breakers — — adjustable, precautions against interference with setting	533-01
— road-warming	554-06		
— segregation from other services	528-02	— electrode water heaters for	554-03
— selection and erection	Chap. 52, Appx. 4	— emergency switching	537-04
— single-core metal-sheathed, bonding	523-05	— miniature, as protection against shock	413-02
— soil-warming	554-06	— nominal current to be indicated	533-01
— space factors	522-08		
— sunlight, exposed to	522-11	— overcurrent settings, precautions against interference with	533-01
— supports, general	521-03, 522-08 Appx. 4 Table 4A		
		— overload protection by	432-02, 432-03
— telecommunication, segregation	528-01	— position of, fundamental requirements	130-03
— temperatures for (See Temperature)		— short-circuit protection by	432-02, 432-04, 434-03
— terminations	Sec. 526	Circuit, definition	p. 8
— thermal insulation, in	523-04	Circuit disconnecting times (See Disconnecting times)	
— trenches, correction factors	Appx. 4	Circuit impedances, for automatic disconnection	413-02
— underground	522-06 Appx. 4	Circuit protective conductor, definition	p. 9
— voltage drop	525-01 Appx. 4	Circuit protective conductor, of ring circuit	543-02
— water or moisture in contact with	522-03	Circuits — — 'Class II', protective conductors in	413-03, 471-09
Capacitive electrical energy, discharge of	461-01	— division of installation into	314-01
Caravan, definition	p. 8	— final (See Final circuits)	
Caravan site, definition	p. 8	— isolation	Sec. 461
Caravan site installations — — applicability of the Regulations to	Sec. 608	— number of	314-01
		— protective conductor impedances for	Table 41C
— periodic inspection and testing	608-07	— ring (See Final circuits, ring)	
— plugs and socket-outlets for	608-08, 608-13	— safety separated, arrangement of	413-06
Caravans —		— safety services — — independence of other circuits	562, 313-02
— bonding	Sec. 608		
— cable supports	Appx. 4		
— connectors	608-08		
— couplers	608-08	— selection and erection	Sec. 563
— inlets	608-07	— SELV, arrangement of	411-02
— luminaires	608-08		

— separation of	314-01, 411-02	stresses	521-03
— switching of	476-03	— environmental conditions	Sec. 522
Classification, safety services sources	Sec. 352	— erection of, fundamental requirement	130-02
Class I equipment, definition	p. 9	— floor-warming	554-06 (Table 55C)
Class II equipment, definition	p. 9	— identification	Sec. 514
Class II equipment or equivalent insulation, protection by —		— insulating of, fundamental requirement	130-02
— application of	471-09	— joints	Sec. 526
— application to complete installation or circuit	471-09	— mechanical stresses	522-06, 07, 08
		— neutral, cross-sectional area	524-02
— conductive parts, connection to protective conductor prohibited	413-03	— operational conditions	Sec. 523
		— overhead (*Overhead lines, Overhead wiring*)	
— enclosures of equipment	413-03	— paralleled	433-03, 434-04, 523-02
— insulating screws not to be relied upon	413-03	— protective (*See Protective conductors*)	
— lids or doors in enclosures	413-03	— resistance under fault conditions	434-03, 543-01
— preferred, for agricultural installations	Sec. 605	— road-warming	554-06
		— selection and erection	Chap. 52
— protective conductors in relation to	413-03, 417-09	— soil-warming	554-06
		— soldering of	522-05
— reinforced insulation during erection	413-03	— supports, general	521-03, 521-08, Appx 4
		— switchboard	523-01, 523-03
— supplementary insulation during erection	413-03	— terminations	Sec. 526
		— voltage drop	Sec. 525
— testing	413-03	— water or moisture exposed to	522-03
— type-tested equipment	413-03		
Clocks, plugs and socket-outlets for	553-01	Conduit and conduit systems —	
		— bends	522-08
CNE conductor (*See PEN conductor*)		— boxes	526-04, 543-02
Codes of Practice	Preface p. viii, Appx. 1	— cable capacities of	522-08
Colour —		— colour identification of	514-02
— emergency switching devices	537-04	— drainage of	522-03
		— erection to be completed before drawing in	522-08-02
— fireman's switch	537-04	— fixing (*See Supports*)	
— identification of cables and conductors	Sec. 514	— flexible, prohibited as protective conductor	543-02
— identification of conduits	514-02	— joints	526-03, 526-04
Combined protective and functional purposes		— metallic —	
— earthing arrangements for,	Sec. 546	— pin-grip and plain slip formed by	543-02
Committee, Wiring Regulations, constitution	p. vii	— socket-outlets in	543-02
Compatibility of equipment	331-01, 312-05	— non-metallic, ambient temperature for	527-01
Compliance with Standards	Sec. 511		
Completion certificate —		— non-metallic, boxes suspending luminaires	527-01
— form of	Appx. 6	— outdoors, on walls	522-06
— provision of	732-01, Ch. 74.	— passing through floors and walls	527-02
Concealed cables in walls or partitions	522-17		
		— prefabricated	522-06
Concrete, steel reinforcement (*See Steel*)		— protective conductor formed by	543-02
Conductive parts (*See also Exposed conductive parts, Extraneous conductive parts*)		— selection of types	521-04
		— short lengths for mechanical protection of cables	471-13
Conductive parts in 'Class II' enclosures	413-03		
		— space factors in	522-08
Conductors —		— supports	522-08
— aluminium (*See Aluminium conductors*)		— terminations	Sect. 526
— ambient temperature	522-01	— underground, cables for	Sect. 522
— bare, cables connected to	523-03	— water or moisture in contact with	522-03
— bare, current-carrying capacity			
— bare, on switchboards	523-03	Confined (restrictive) conductive location —	
— bonding (*See Bonding conductors*)		— definition	p. 14
— copperclad (*See Aluminium conductors*)		— earth fault loop impedance in	471-08
— corrosive or polluting substances, exposure to	522-05	— nominal voltage of SELV circuits in	471-02
— colour identification of	514-06		
— current-carrying capacity	Appx. 4, Sec. 523	Connection, electrical, fundamental requirements	130-02
— electromechanical			

Connector, definition	p. 9
Connectors, caravan	608-07
Connectors, clock	553-01
Construction of equipment, applicability of the Regulations to	110-04
Construction (General Provisions) Regulations	Appx. 2
Construction sites	Sec. 604
– cable couplers	553-02
– periodic inspection and testing	Appx. 6
– statutory regulations	Appx. 2; 120-02
Contactors, emergency switching by, to fail safe	537-04
Continuity tests	713-02
Contracts, citation of the Regulations in	120-02
Controlgear, labels and indicators for	514-01
Controls in bathrooms	601-08
Cord-operated switches in bathrooms	601-08
Correction factors –	
– ambient temperature for cables	Appx. 4
– cables in trenches	Appx. 4
– grouping of cables	Appx. 4
Corrosion –	
– cables, general	522-05
– earth electrodes	522-03
– earthing conductors	522-05
– heating cables	554-06
Corrosive atmospheres, equipment exposed to	522-05
Corrosive substances, wiring systems exposed to	522-05
Coupler, cable (See Cable coupler)	
Coupler, caravan	Sect. 608
Creosote, in contact with non-metallic wiring systems	522-04
Current –	
– nature of assessment	311, 313-01
– nominal, of fuses and circuit-breakers	533
– suitability of equipment for	512-02
Current-carrying capacity –	
– cables and conductors	130-02, Sec. 523
– definition	p. 9
– parallel conductors	433-03
Current transformers, omissions of overload protection	473-02
Current-using equipment –	
– definition	p. 9
– power demand	130-02
– selection and erection of	554

D

Danger, definition	p. 9
Danger notices (See Warning notices)	
Data processing equipment	607-01
Data transmission circuits, applicability of the Regulations to	110-01
D.C. feedback, assessment for compatibility	331-01
D.C. plugs and socket-outlets, prohibited for switching	537-05
Defects or omissions, on test and inspection	Ch 74
Defined conditions, for cable current-carrying capacities	Appx. 4 (Table 4A)
Definitions	Part 2
Demand, maximum, assessment of	311-01
Demand, maximum, suitability of supply for	313-01
Departures from the Regulations	Chap. 74
Design current, definition	p. 9
Diagrams, provision of	514-09
Diagrams, availability for inspection and testing	711-01
Dielectric, flammable liquid	422-05
Direct and indirect contact, protection against both –	471-02, 471-03
– application of	410-01, 471-02
– cables as	471-09
– in areas accessible only to skilled or instructed persons	471-13
Direct contact, definition	p. 9
Direct contact, protection against –	412, 471
– application of	471-04 to 471-07
– in areas accessible only to skilled or instructed persons	471-13
– supplementary, by residual current device	471-08, 471-16
– testing of	713-05, 713-07
Discharge lighting –	
– applicability of the Regulations to	110-01
– circuits, mineral-insulated cables in	521-01
– circuits, reduced neutral prohibited	542-02
– fireman's switch	537-04
– high voltage (See High voltage)	
– isolation	476-02
Discharge of energy, for isolation	461-01
Discharge of energy, protection by limitation of	411-04, 471-03
Disconnecting times –	
– complying with temperature limits for protective conductors	Appx. 3
– equipment in bathrooms	Sec. 601
– reduced low voltage circuits	471-15
– shock protection, fixed equipment outside main equipotential zone	471-08
– shock protection, general	413-02
Discrimination	
– overcurrent protective devices	533-01
– residual current devices	531-02
Disputes with supply undertakings	Appx. 2
Distribution boards, backless, erection of	422-01
Diversity, assessment of	311-01
Doors in 'Class II' enclosures	413-03
Double insulation, definition	p. 9
Double insulation, of equipment	413-03
Drainage of conduit systems	522-03
Duct, definition	p. 9
Duct (See also Cable ducts)	
Ducting (See Cable ducting)	
Dust conditions, onerous	522-04

Dynamic stresses, for short-circuit protective devices	434-03

E

Earth —	
— connections to	331-01, 542
— definition	p. 9
— protection of persons in contact with general mass of	471-08
Earth electrode —	
— caravan sites	Sec. 608
— definition	p. 10
— IT systems	413-02
— resistance —	
— allowance for soil drying and freezing	542-02
— allowance for corrosion	542-02
— area, definition	p. 21
— definition	p. 14
— for protection against shock	413-02
— test of	713-11
Earth electrodes —	
— electric fences	605-14
— selection of	542-02
Earth fault current —	
— adequacy of earthing arrangements for	542-01, 543-01
— fundamental requirements	130-04
Earth fault loop impedance —	
— agricultural installations	Sec. 605
— body resistance, in relation to	471-08
— definition	p. 10
— equipment used outside main equipotential zone	471-08
— external to installation, assessment	313-01
— for automatic disconnection for protection against indirect contact	413-02
— measurement	Sec. 713
— reduced low voltage circuits	471-15
— testing	713-10
Earth-free local equipotential bonding, protection by —	413-05
— application of	471-11
— avoidance of contact with Earth	413-05
— bonding of parts	413-05
— limited to special situations	471-11
— precautions at entrance to location	413-05
Earth-free location (See Non-conducting location)	
Earth leakage current —	
— assessment for compatibility	331-01
— data processing equipment	607-02, 607-05
— definition	p. 10
— fundamental requirements	130-04
— HV electrode water heaters and boilers	554-04
— residual current devices and	531-02
suitability of earthing arrangements for	542-01
Earth loop impedance (See Earth fault loop impedance)	
Earth monitoring	543-03
Earth concentric wiring, definition	p. 10
Earth concentric wiring (See also PEN conductors)	
Earthed equipotential bonding and automatic disconnection, protection by —	413-02, 471-08
— agricultural installations	Sec. 605
— applications of	471-08
— basic requirements	413-02
— body resistance in relation to	471-08
— caravan installations	471- Sec. 608
— coordination of characteristics	413-02
— devices for, types	413-02
— disconnecting times	413-02, 471-08 471-15
— earth fault loop impedances	413-02
— equipment used outside main equipotential zone	Sec. 471
— exposed-conductive-parts, connections of —	
— IT systems	413-02
— TN systems	413-02
— TT systems	413-02
— local supplementary bonding	413-02
— main equipotential bonding	413-02
— reduced system voltages and	471-15
— residual current device for	471-08
— socket-outlet circuits in TT systems	471-08
Earthing —	
— arrangements —	
— determination of type of	312-03
— selection and erection	Chap. 54
— combined protective and functional purposes	546-01
— conductors, electrode boilers	554-03
— conductors, selection and erection	542-03
— connections, warning notices at	514-13
— exposed-conductive-parts —	
— IT systems	413-02
— TN systems	413-02
— TT systems	413-02
— high leakage	Sec. 607
— lightning protection systems	Sec. 541
— neutral of supply	546-02, 554-03
— prohibited, in earth-free local bonded location	413-05
— prohibited, in non-conducting location	413-04
— resistance —	
— provision for measurement of	542-04
— selection of value of	542-01
— variations in	531-02
— terminal, main —	
— connection to Earth	531-02, 542-01
— definition	p. 12
— selection and erection	542-04
Editions, list of	p. vi
Effects of the Regulations	Chap. 12
Electric braking, with emergency stopping	537-04

Electric shock —	
— definition	p. 10
— in case of fault *(See Indirect contact)*	
— in normal service *(see Direct contact)*	
— protection against	Chap. 41
— current *(See Shock current)*	
safety services	564-01
Electric surface heating systems	554-07
Electric traction equipment, the Regulations not applicable to	110-02
Electrical engineer *(See Engineer)*	
Electrical equipment, definition	p. 10
Electrical equipment *(See also Equipment)*	
Electrical installation, definition	p. 10
Electrical separation, protection by —	413-06
— application of	471-12
— bonding of equipment	413-06
— exposed metalwork of circuit	413-06
— flexible cables for	413-06
— for one item of equipment	413-06
— for several items of equipment	413-06, 413-06 471-12
— protective conductors for	413-06
— separation of circuit	413-06
— socket-outlets for	413-06
— supplies for	413-06
— testing of	713-06
— voltage limitation	413-06
Electricity at Work Regulations 1989 —	
— applicable to construction sites	Appx. 2
— areas accessible to skilled persons	471-13
— passageways and working platforms	471-13
— relationship of the Regulations to	120-02, Appx. 2
— requirements for enclosures	412-03
Electricity Supply Regulations —	
— relationship of the Regulations to	120-02, Appx. 2
— selection and erection of	554-03
Electrodynamic effects *(See also Electromechanical, Electromagnetic)*	
Electrodynamic effects on protective conductors	543-03
Electrolysis, precautions against —	
— earthing arrangements	543-02
— wiring system metalwork	
Electromagnetic effects in cables	521-02
Electromechanical stresses —	
— conductors and cables	521-03
— earthing arrangements	542-07
— protective conductors	543-03
Electronic devices — *(See also Semiconductors)*	
— as safety source for SELV circuit	411-02
Emergency lighting systems —	
— segregation from LV circuits	528-01
— supplies for	313-02
Emergency stopping	463-01, 537-04
Emergency switching —	
— application of	476-03
— definition	p. 10
— provision of	476-03
— safety services	476-03
— selection of devices for	537-04
Enclosure, definition	p. 10
Enclosures —	
— effect on accessibility	Sec. 513
— joints and terminations for	526-02
— protective conductors formed by	543-02
— provided during erection, testing of	713-07
Enclosures, protection by —	
— Class II or equivalent	413-03
— degrees of protection	412-03
— fire-resistant, for fixed equipment	422-01
— lids or doors in	413-03
— openings in	412-03
— provided during erection, testing of	713-07
— removal or opening of	412-03
— securing of	412-03, 413-03
Energy, limitation of discharge of	411-04
Engineer, suitably qualified electrical	120-01, 521-07 Appx. 6
Environmental conditions, of cables and conductors	Sec. 522
Environmental conditions *(See External influences)*	
Equipment —	
— accessibility	130-07, 513
— agricultural installations	Sec 605
— applicability of the Regulations to	110-04
— Class II	413-03, 471-09
— compatibility of	512-05
— construction of	Sec. 511
— containing flammable dielectric liquid	422-05
— current-using, definition	p. 9
— current-using, requirements	Sec. 554
— data processing equipment	607-01
— definition	p. 10
— electrically heated, mechanical maintenance switching for	476-02
— erection of	Part 5
— fixed —	
— definition	p. 11
— disconnecting times for	413-02
— earth fault loop impedances for	413-02
— for safety services supplies	562-01
— heat dissipation	422-01
— fundamental requirements	Sec. 511
— guarding, against burns	423-01
— having exposed live parts	471-13
— heating, mechanical maintenance switching	476-02
— incorporating motors	552-01
— isolation and switching of	Chap.46
— maintenance of	Sec. 462
— mechanical maintenance switching for	462-01

— mobile, in non-conducting location	413-04
— oil-filled	422-01
— operational conditions	Sec. 512
— outdoors, circuits supplying	471-08
— portable —	
— connection of	553-01-07
— definition	p. 13
— in non-conducting location	413-04
— selection of	Part 5
— stationary, definition	p. 14
— suitability for maximum power demand	300-01
— surface temperature of	422-01, 423-01
— testing of	Part 6
— type-tested	413-03
— used outside main equipotential zone	471-08
Equipotential bonding —	
— caravans	Sec. 608
— conductors —	
— main	413-02, Sec. 608
— selection and erection of	547-02
— supplementary	413-02, 547-03
— definition	p. 10
— earth-free local	413-05, 471-11
— local supplementary, bathrooms	Sec. 601
— local supplementary, general	413-02
— main	413-02
- safety separated circuits, in	413-06
— testing of	713-02
— zone, equipment used outside	471-08
ERA Technology Ltd.	Appx. 4
Erection of equipment	Part. 5
Established materials and methods, use of	120-04
European Committee for Electrotechnical Standardization (CENELEC)	Preface p. viii
Exciter circuits of machines, omission of overload protection	473-01
Exhibition buildings, discharge lighting in	476-03
Exclusions from Regulations	110-02
Explosive atmospheres, installations in —	
— applicability of the Regulations to	110-01
- Statutory regulations for	Appx. 2
Exposed-conductive-parts —	
— bathrooms	Sec. 601
— connection, for earthing and automatic disconnection —	
— IT systems	413-02
— TN systems	413-02
— TT systems	413-02
— earth-free local bonded location	413-05
— exemptions from protection against indirect contact	471-13
— FELV systems	411-03
— local bonding of	413-02
— protective conductors formed by	543-02
— SELV circuits	411-02
Extensions to installations	130-09
External influence,	Chap. 72
concise list	Sec. 522, Appx. 5
definition	p. 11
— earthing arrangements to be suitable for	542-01
— equipment to be suitable for	Sec. 512
Extra-low voltage—(See also Functional and SELV)	110-03
— cables for	411-02
— definition	p. 15
— range	p. 15
Extraneous-conductive-part, definition	p. 11
— bathrooms	Sec. 601
— bonding of	413-02
— caravans	Sec. 608
— connection to Earth —	
— IT systems	413-02
— propagating potential outside non-conducting location	413-04
— protective conductors formed by	543-02
	713-02
— SELV circuits	411-02

F

Factory-built assembly (of LV switchgear) —	
— enclosure used as protective conductor	543-02
— having total insulation	413-03
Factory Regulations (See Electricity (Factories Act) Special Regulations)	
Farms (See Agricultural installations)	
Fauna, damage to cables by	522-10
fence controllers, as example of limitation of discharge of energy	471-03
Fences, electric, erection of	605-14
Final circuit, definition	p. 11
Final circuits —	
— arrangements of, general	314-01
— connection to distribution board	314-01
— control of	314-01
— disconnecting times, for earthing and automatic disconnection (See Disconnecting times)	
— earth fault loop impedances (See Earth fault loop impedances)	
— number of points supplied	314-01
— number required	314-01
— ring —	
— continuity test	Sec. 713
— definition	p. 14
— protective conductor of	543-02
— separate control of	314-01
— separation of	314-01
— standard arrangements of	314-01
Fire alarm systems, applicability of the Regulations to	110-01
Fire alarm systems, segregation from LV circuits	528-01
Fire authority, siting of fireman's switch	476-03

Fire barriers	527-02
Fire conditions, safety services for	561-01
Fire, protection against —	
— appliances, fixed	422-01
— distribution boards, backless	422-01
— equipment containing flammable liquid	422-01
— fire hazard to building materials	422-01
— fire-resistant shield or enclosure	422-01
— fundamental requirement	130-08
— heat dissipation of equipment	422-01
— heating cables	Sec. 554
— lamps	422-01
— luminaires	422-01
— surface temperature of equipment	422-01
— ventilation or equipment	422-01
Fireman's switch, provision of	476-03
Fireman's switch, selection of	537-04
Fire-resistant —	
— shields or enclosures for equipment	422-01
— structural elements, cables passing through	527-02
Fire risks, special, the Regulations not applicable to	110-02
Fixed equipment —	
— definition	p. 11
— disconnecting times for	413-02
— earth fault loop impedances for	413-02
— heat dissipation	422-01
Flammable dielectric liquid, in equipment	422-01
Flexible cables and cords (See Cables, Flexible)	
Flexible conduits, prohibited as protective conductors	543-02
Floor-warming systems —	
— cables and conductors for	554-06
Floors —	
— cables passing through	527-02
— cables under	522-17
— conductive, in earth-free bonded location	413-05
— insulating, resistance of	413-04
Fluctuating loads, assessment for compatibility	331-01
Fluorescent lighting loads, assessment for compatibility of harmonics	331-01
Fluorescent lighting (See also Discharge lighting)	
Foreign Standards	Preface p. viii
Foundation earthing	542-02
Frequency —	
— effect on current-carrying capacity of conductors	Appx. 4
— suitability of equipment for	512-03
— supply, assessment of	313-01
Functional earthing —	
— assessment of need for	331-01
— definition	p. 11
— protective earthing in relation to	542-01
Functional extra-low voltage definition	p. 11, 411-03, 471-14
Functional switching, devices for	537-05
Fundamental requirements for safety	Chap. 13
Fuse element, definition	p. 11
Fuse elements, for semi-enclosed fuses	533-01
Fuse link, definition	p. 11
Fuse links —	
— marking of intended type	533-01
— non-interchangeability of	533-01
— replaceable by unskilled persons	533-01
Fused —	
— plugs	533-01, 553-01
Fuses — (See also Overcurrent protective devices)	
— accessible to unskilled persons	533-01
— cartridge type, preferred	533-01
— in plugs	553-01
— markings of	533-01
— nominal current to be indicated	533-01
— non-interchangeability of links	533-01
— overload and short-circuit protection by	432-02, 533-01
— prohibited in earthed neutral	130-05
— removable whilst energised	533-01
— selection of	533-01
— semi-enclosed —	
— elements for	533-01
— overload protection by	433-02, Appx. 4
— special ambient temperature correction factors for cables protected by	Appx. 4
— shock protection by	413-02
— short-circuit protection by	432-04
— voltage rating for fault conditions	530-01
Fusing factor	533-01

G

Gas installation pipe, definition	p. 11
Gas meters	547-02
Gas pipes —	
— bonding at points of contact with	528-02
— main bonding of	413-02, 547-01
— public, prohibited as earth electrodes	542-02
— segregation from	528-02
General characteristics of installation, assessment of	300-01
Glossary (See Definitions)	
Guarding of equipment, against burns	423-01
Guarding of lamps and luminaires	422-01

H

Harmonic currents, assessment for compatibility	331-01
Harmonization Documents	Preface p. viii
Health and Safety at Work etc. Act	Appx. 2
Heat dissipation of equipment	422-01
heating appliances —	
— bathrooms	Sec. 601
Heating cables	554-06
Heating equipment, mechanical maintenance switching for	476-02
High-frequency oscillations, assessment for compatibility	331-01
High voltage discharge lighting	554-02
— fireman's switch	476-03
High voltage electrode water heaters and boilers	554-03
Highly Inflammable Liquids and Liquified Petroleum Gases Regulations	Appx. 2
Hoist shaft, cables in	528-02
Household or similar installations —	
— socket-outlet circuits in TT systems	471-08
— socket-outlets for, selection	553-01
Hydrocarbons, exposure of non-metallic wiring system to	522-05

I

Immersion heaters *(See also Water heaters)*	
Impedance —	
— circuit, for automatic disconnection	413-02
— earth fault loop *(See Earth fault loop impedance)*	
— earthing, in IT systems	413-02
—'protective' *(See Limitation of discharge of energy)*	
Indicators —	
— circuit-breaker overcurrent settings	533-01
— isolation devices, operation of	537-02
— mechanical maintenance switching	537-03
— switchgear, operation of remote	Sec. 514
Indirect contact, definition	p. 12
Indirect contact, protection against —	410-01, Secs. 413, 471
— application of	Sec. 471, Part 6
— automatic disconnection of supply	413-02, 471-08
— Class II equipment or equivalent	413-03, 471-09
— earth-free local equipotential bonding	413-05, 471-11
— earthed equipotential bonding and automatic disconnection	413-02, 471-08
— electrical separation	413-06, 471-12
— exemptions, special	471-13
— non-conducting location	413, 471-10
— provision of	Sec. 413
— testing	Sec. 713
Inductance, mutual, assessment for compatibility	331-01
Industrial plugs and socket-outlets	553-01
Inlet, caravan	Sec. 608
Inspection —	
— certificate	Appx. 6
— equipment to be accessible for	130-02
— initial, check list	Chap. 71
— initial, requirement for	Sec. 712
— installation, fundamental requirement	130-10
—periodic —	
— assessment of need for	341-01
— lengths of periods	Appx. 6 (footnotes)
— notice on	514-12
— requirement for	Chap. 63
— caravans	608-07
— visual, requirement for	Sec. 712
Installation, electrical, definition	p. 10
'Instantaneous' water heaters in bathrooms	Sec. 601
'Instantaneous' water heaters *(See also Water heaters)*	
Instructed person, definition	p. 11
Instructed persons —	
— areas reserved for, to be indicated by signs	471-13
— emergency switching of safety services reserved for	476-03
— obstacles permissible for protection of	471-06, 471-13
— placing out of reach permissible for protection of	471-06, 471-13
— safety services sources	Sec. 562
Insulating —	
— applied during erection	412-02, Sec. 713
— basic, definition	p. 7
— double, definition	p. 9
— double, of equipment	413-03
— equivalent to Class II *(See Class II equipment)*	
— joints in conductors	526-02
— live parts	412-01, 471-94
— monitoring, in IT systems	413-02
— non-conducting floors walls	413-04, 713
— PEN conductors	546-02
— protection by insulation of live parts	412-01, 471-04
— protective conductors	543-03
— reinforced —	
— applied during erection	413-18 (v)
— definition	p. 14
— of equipment	413-03
— resistance, testing	Sec. 713
— SELV circuits	411-02
— site-built assemblies	412-02, 413-03 471-09, Sec. 713
— supplementary, applied during erection	413-03, 471-09 713-05
— supplementary, definition	p. 14
— thermal, cables in	523-04
Interlocking of enclosure opening	412-03
Interlocking of isolating devices	461-01

International Electrotechnical Commission	Preface p. viii, Chap. 32
Intruder alarm systems, applicability of the Regulations to	110-01
Isolation —	Secs. 460 & 461 476-02
— at origin of installation	476-02
— definition	p. 12
— devices for, contact distances	537-02
— devices for, selection and erection	Sec. 537
— discharge lighting	476-02
— fundamental requirements	130-06
— indication of	537-02
— main, at origin of installation	476-02
— motor circuits	476-02
— motors, fundamental requirements	130-06
— multipole devices preferred	537-02
— off-load, securing against inadvertent operation	537-02
— PEN conductor not to be isolated	460-01, 546-02
— prevention of inadvertent reclosure	537-02
— remote devices for	476-02
— semiconductors unsuitable for	537-02
— single-pole devices for	537-02
I.P. protection	412-03
Isolator —	
— contact distances	537-02
— for maintenance of main switchgear	476-02
— off-load, securing against inadvertent operation	537-02
— semiconductors unsuitable as	537-02
IT system —	
—autotransformers, step-up, prohibited	551-01
— connection of installation to earth	542-04
— definition	p. 15
— distribution of neutral in	473-03, 512-01
— earthing of live parts in	413-02
— earthing resistance	542-01
— exposed-conductive-parts in, connection of	413-02
— first fault in	413-02
— insulation monitoring devices for	413-02
— neutral distribution in	473-03
— overload protection	473-01
— overvoltage in	413-02
— shock protective devices in	413-02
— variation of earthing resistance	531-02
— voltage oscillations in	413-02

J

Joints —	
— cable, accessibility of	Sec. 513
— cable and conductor	Sec. 527
— conduit system, accessibility of	Sec. 513
— earthing arrangements	542-04
— fundamental requirements	130-02
— protective conductor	543-03
Joists, cables passing	522-06

K

k values —	
— in relation to protective conductors	Tables 54B, 54C, 54D and 54E, 54F, 54G
— in relation to short-circuit currents	Table 43A
Keys —	
— adjustment of overcurrent settings by	533-01
— areas accessible only to skilled persons by	471-13
— opening enclosures by	412-03, 413-03

L

Labels — (See also Marking, Warning notices)	
— earth electrode connections	542-03, 514-13
— switchgear and controlgear	514-09
Lacquers, as insulation	412-02 413-03
Lampholders —	
— as example of permissible opening in enclosure	471-05
— bathrooms	Sec. 601
— batten, defined as luminaire	p. 12
— centre contact	553-03
— Edison type screw	553-03
— filament lamp, voltage limit	553-03
— overcurrent protection of	553-03
— pendant, defined as luminaire	p. 12
— selection and erection of	553-03
— temperature rating	553-03
Lamps, fire hazard from	422-01
Leakage currents and residual current devices	531-02
Leakage currents (See also Earth leakage)	
Licensing, premises subject to	120-03
Lids in 'Class II' enclosures	413-03
Life of installation, assessment of	341-01
Lift shafts, cables in	528-02
Lifting magnets, omission of overload protection	473-01
Lightning protection, the Regulations not applicable to	110-02
Lightning protection earthing for	541-01
Limitation of discharge of energy, protection by	411-04, 471-03
Live conductors, assessment of types	312-02
Live conductors, earthing of, prohibited in IT systems	413-02
Live part, definition	p. 12
Live parts —	
— bare, in SELV circuits	411-02

— bare, placing out of reach	412-05
— of SELV circuits, separation of	411-02
Livestock —	
— protective measures in situations for	Sec. 605
— wiring systems in locations for	522-01
Local supplementary bonding	413-02, 471-15
Locations exposed to fire risk, safety services in	563-01
Low voltage —	
— range	110-03
— reduced	471-15
Luminaire, definition	p. 12
Luminaire track system (See definition of Socket-outlet)	p. 14
Luminaires —	
— bathrooms	601-11
— caravans	608-08
— fire hazard from	422-01
— ceiling rose for	553-04
— mass suspended	Appx. 4, 521-03 522-08, 544-01
— portable, connection of	553-01
— supported by flexible cords	522-02
— suspended from non-metallic boxes	527-01
— switches mounted on	476-03
— switching for	476-03

M

Machines — (See also Motors)	
— emergency switching for	476-03
— rotating, exciter circuits of	473-01
— rotating, selection and erection of	Sec. 552
Magnetic —	
— circuit of residual current device	531-02
— equipment, mechanical maintenance switching for	Sec. 462, 537-03
— fields, effect on residual current devices	531-02
Magnets, lifting, omission of overload protection	473-02
Main earthing terminal —	
— connection to earth	542-01
— definition	p. 12
— selection and erection of	542-04
Main equipotential bonding —	
— conductors, selection and erection	547-02
— provision of	413-02
— zone, equipment used outside	471-08
Main switch for installation	460-01-02
Maintainability, assessment of	341-01
Maintenance of equipment —	
— assessment of frequency of	341-01
— fundamental requirement	130-02
Markets, closed, discharge lighting in	476-03
Marking — (See also Labels, Warning notices)	
— buried cables	522-06
— caravan inlets	608-07
— emergency switching devices	463-01
— fuses and circuit-breakers	533-01
— isolating devices	461-01
— mechanical maintenance switching devices	462-01
Mass supported from non-metallic boxes	521-03, 522-08 544-01
Materials —	
— established, use of	120-04
— new, use of	120-05
— proper, use of	130-01
Maximum demand, assessment of	311-01
Maximum demand, suitability of supply for	313-01
Mechanical maintenance, definition	p. 12
Mechanical maintenance switching, requirements	Sec. 462, 537-03
Mechanical maintenance switching devices, selection of	537-03
Mechanical stresses —	
— cables and conductors	434-01, 521-02, 522-06
— earthing arrangements	542-01, 542-03
— earthing conductors	542-03
— heating cables and conductors	554-06
— plugs and socket-outlets	553-01
— protective conductors	543-01, 543-03,
— short-circuit	434-01
Metalwork — (See also Exposed-conductive-parts, Extraneous-conductive-parts)	
— bonding of	413-02
— cables passing through	522-06, 522-08
— exposed, of safety separated circuits	413-06
— of other public services, prohibited as earth electrode	542-02
— wiring system, as protective conductor	543-02
— wiring system, protection against corrosion	522-05
Mines and quarries, installations at —	
— statutory regulations for	Appx. 2
— the Regulations not applicable to	110-02
Miniature circuit-breakers (See Circuit-breakers)	
Mobile equipment in non-conducting location	413-04
Mobile safety sources —	
— electrically separated circuits supplied by	413-06
— SELV circuits supplied by	411-02
Monitoring systems, earthing	543-03
Monitoring systems, insulation, IT systems	413-02
Motor vehicles, applicability of the Regulations to	110-02
Motors —	
— automatic restarting	552-01
— control of	552-01
— disconnection, fundamental requirement	130-05, 130-06
— frequent start	Sec. 552
— isolation of	476-02

— mechanical maintenance switching	462-01, 537-03
— stalled, capability of emergency switching for	537-04
— starters, autotransformers	551-01
— starters, coordination of overload and short-circuit protection	435-01
— starting currents	552-01
— starting, voltage drop	525-01
Mutual detrimental influence —	
— different types of equipment	Sec. 515
— electrical and other services	Sec. 515
— LV circuits and other circuits	528-01
— protective measures	470-01
Mutual inductance, assessment for compatibility	331-01

N

Neutral —	
— conductor —	
— combined with protective conductor *(See PEN conductors)*	p. 12
— definition	p. 12
— earthing of	130-05, 471-15-04, 546-02, 554-03 Appx. 2
— fuse prohibited in	130-05, 530-01
— isolation of	460-01
— minimum cross-sectional area	524-02
— overcurrent detection in	473-03
— reduced, prohibited in discharge lighting circuits	524-02
— single-pole devices prohibited in	130-05, 530-01
— switching of	530-01
— distribution of, in IT systems	473-03
— earthing of	130-05, 471-15-04, 546-02, 554-03, Appx. 2
— links	476-03, 537-02
— reduced, prohibited in discharge lighting circuits	524-02
New materials and methods, use of	120-04
Nominal current of fuses and circuit-breakers	533-01
Nominal voltage —	
— assessment	313-01
— definition	p. 15
— reduced low voltage circuits	471-15
— SELV circuits	411-02, Sec. 605
Non-automatic sources for safety services	352-01
Non-conducting location, protection by —	413-04, 471-10
— application of	471-10
— basic requirements	413-04
— limited to special situations	471-10
— mobile and portable equipment in	413-04
— not recognised for general use	471-10
— permanency of arrangements	413-04
— precautions against propagations of potential outside the location	413-04
— protective conductors prohibited	413-04
— resistance of floors and walls	413-04, Sec. 713
— socket-outlets in	413-04
Notices —	Sec. 514
— fireman's switch	537-04
— periodic inspection and testing	514-12
— isolation	514-11
— Warning —	
— caravans	608-07
— earthing and bonding conductors	514-13
— emergency switching	537-04
— voltage	514-10
Numbering system of Regulations	preface p. ix

O

Object and effects of the Regulations	Chap. 12
Obstacle, definition	p. 12
Obstacles, protection by —	412-04, 471-06
— application of	471-06
— basic requirements	412-04
— limited to areas for skilled or instructed persons only	471-06, 471-13
Offshore installations, Regulations not applicable to	110-02
Oil —	
— exposure of non-metallic wiring systems to	522-05
— filled equipment	422-01
— services, bonding at points of contact	528-02
— services, segregation from	528-02
Operational conditions, cables and conductors	Sec. 512
Operational conditions, equipment	Sec. 512
Origin of an installation —	
— definition	p. 12
— isolation at	476-01
Oscillations, HF, assessment for compatibility	331-01
Oscillations, voltage in IT systems	413-02
Other services, bonding of	130-04
Overcurrent, definition	p. 12
Overcurrent detection —	
— neutral conductor, IT systems	473-03
— neutral conductor, TN or TT systems	473-03
— phase conductors	473-03
Overcurrent, limitation by supply characteristics	436-01
Overcurrent, protection against —	Chap. 43, Sec. 473
— at origin of installation	130-03
— basic requirement	130-03, 431-01
— discrimination in	533-01
— electrode water heaters and boilers	554-03

— fundamental requirement	130-03
— lampholders	553-03
— limitation by supply characteristics	436-01
— motors	552-01
— neutral conductor, IT systems	473-03
— neutral conductor, TN or TT systems	473-03
— paralleled conductors	433-03, 434-04
— phase conductors	473-03
— settings of adjustable circuit-breakers	533-01
Overcurrent protective devices —	
— at origin, assessment of	313-01
— at origin, use of supply undertaking device	130-05
— co-ordination of characteristics, for overload and short-circuit protection	435-01
— discrimination	533-01
— fundamental requirements	130-05
— overload and short-circuit protection	432-02
— overload protection only	432-03
— safety services	563-01
— selection and erection of	Sec. 533
— shock protection, as	413-02
— short-circuit protection only	432-04
— supply undertaking's, use of	130-05
Overcurrent settings of circuit-breakers	533-01
Overhead lines —	
— insulator wall brackets, exemption from protection against indirect contact	471-13
— placing out of reach	412-05
— types of conductor	521-01
Overheating of switchgear	422-01
Overload —	
— current, definition	p. 12
— protection against —	
— application of	473-01
— coordination with short-circuit protection	435-01
— omission of	473-01
— omission, for safety services	563-01
— paralleled conductors	433-03
— position of devices for	473-01
— protective devices	432-01, 473-01, 432-03, 433-01
Overvoltages, in IT systems	413-02
Overvoltages, transient, assessment for compatibility	331-01

P

Paints, as insulation	413-03
Paralleled cables, current-carrying capacity	523-02
Paralleled conductors, overload protection	433-03
Paralleled conductors, short-circuit protection	434-04
Parallel operation, safety services supplies	Sec. 566
Partitions, cables in,	522-17
Partitions in trunking	528-01
Passageways for open type switchboards	471-13
PEN conductor, definition	p. 12
PEN conductors —	
— cable enclosure prohibited as	543-02
— isolation or switching of, prohibited	460-01
— residual current device with, prohibited	413-02
— selection and erection of	Sec. 546
Pendant lampholder, defined as luminaire	p. 12
Pendant luminaire —	
— caravans	608-08
— ceilings rose for	553-04
— mass suspended	554-01
Periodic inspection —	
— assessment of need for	Sec. 732, 130-10
— length of periods	Appx. 6
— notice on	514-12
Persons—	
— instructed, definition	p. 11
— skilled, definition	p. 14
— skilled or instructed, obstacles permissible for protection of	471-06, 471-13
— skilled or instructed, placing out of reach permissible for protection of	471-07, 471-13
— untrained, the Regulations not intended for	120-01
Petroleum (Consolidation) Act	Appx. 2
Phase conductor —	
— definition	p. 12
— identification of	Sec. 514
— overcurrent detection in	473-03
Phase, loss of	474-03
— gas (See Gas pipes)	
— earth electrodes formed by	542-02
— underground, for cables	522-06
— water (See Water pipes)	
Placing out of reach, protection by —	412-05, 471-07
— application of	471-07
— bare live parts	412-02, 412-03
— limited to locations for skilled or instructed persons only	471-07, 471-13
— overhead lines	412-05
— with obstacles	412-05
— zone of accessibility	412-05
Plan of the 16th Edition, Notes on	p. ix
Platforms, working, for open type switchboards	471-13
Plug, definition	p. 13
Plugs —	
— caravans	608-08
— clock	553-01
— construction sites	604-12
— emergency switching by, prohibited	537-04
— FELV circuits	411-03
— functional switching by	537-05
— fused, selection of	553-01
— 'instantaneous' water heaters not to be supplied by	554-05

259

— reduced low voltage circuits	471-15
— selection and erection of	553-01
— SELV circuits	411-02
— shavers	601-08
— special circuits	553-01
PME —	
— Approval,	413-02, Appx. 2
— on caravan sites	608-13
— main bonding connections for	547-02
Point (in wiring), definition	p. 13
Polarity, test of	713-09
Poles, concrete, exemption from protection against indirect contact	471-13
Polluting substances, wiring systems exposed to	522-05
Portable equipment—	
— connection of	553-01-07
— definition	p. 13
— in non-conducting location	413-04
— outdoors	471-16
— prohibited in bathrooms	601-10
Potentially explosive atmospheres, installations in, extent covered by Regulations	110-02
Power demand, suitability of equipment for	130-03, 512-04
Prefabricated equipment, applicability of the Regulations to	110-04
Premises subject to licensing	120-03
Proper materials, use of	130-01
Prospective short-circuit current—	
— at origin, assessment of	313-01
— at points of protective devices	Sec. 432
— determination of	434-02
Protection against electric shock—	
— application of measures	Sec. 471
— protective measures	Chap. 41
— safety services	564-07
Protection against overcurrent—	Chap. 43
Protection against thermal effects	Chap. 42
Protection against electric shock	Chap. 41
Protection against undervoltage	Chap. 45
Protection for safety	Part 4
Protection, supplementary, by residual current device	412-06
Protective conductor, definition	p. 13
Protective conductors—	
— accessibility of connections	543-03
— between separate installations	542-01
bonding, selection and erection of	Sec. 547
— caravan sites	608-13
— 'Class II' equipment in relation to	413-03, 471-09
— colour identification of	514-03
— combined with neutral conductors (See PEN)	
— continuity of	543-03

— cross-sectional areas	543-01 Sec. 547. 546-02
— identification of	514-03
— insulation of	543-03
— non-conducting location, prohibited in	413-04
— preservation of continuity	543-03
— residual current device, to be outside magnetic circuit of	531-02
— ring final circuit	543-02
— safety separated circuits	413-06
— selection and erection of	Secs. 543 to 547
— separate	543-02
— switching prohibited in	543-03
— 10 mm^2 or less	543-02
— testing of	713-02
— types of, description	543-02, 547-03
Protective devices and switches, position of	130-05
Protective devices and switches, identification of	514-01, 514-03
Protective devices (See also Overcurrent protective devices, Residual current devices, Monitoring systems, Undervoltage protective devices)	
'Protective impedance' (See Limitation of discharge of energy)	
Protective Multiple Earthing (See PME)	
Public supply —	
compatibility with	512-05
— IT system prohibited for (in UK)	Fig. 7
— switching arrangements for safety and standby supplies	313-02
— the Regulations not applicable to	110-02

Q

Quarry installations, exclusion	110-02

R

Radio interference suppression equipment, applicability of the Regulations to	110-02
Rapidly fluctuating loads, assessment for compatibility	331-01
Reduced low voltage systems	471-15
Regulations, Statutory—	
— list of	Appx. 2
— relationship with the Regulations	Appx. 2
Regulations, The—	
— citation in contracts	120-01
— departures from	120-05
— editions, list of	preface p. vi
— effect of	120-02
— exclusions from scope of	110-02
— not intended to instruct untrained persons	120-01
— not to be regarded as specification	120-01
— objects	Chap. 12
— Parts 3 to 7, relationship to Chapter 13	120-02
— relationship to statutory regulations	120-02

— scope of	Chap. 11
— voltage ranges dealt with	110-03
Reinforced concrete *(See Steel)*	
Reinforced insulation, definition	p. 14
Reinforced insulation, of equipment	413-03
Reliability of equipment for intended life	341-01
Remote devices for isolation	476-02
Remote switching for mechanical maintenance	537-04
Repairs, notification of need for	743-01
Residual current device, definition	p. 14
Residual current devices—	
— application of, as shock protection	412-06
— auxiliary supply for	531-02
— caravan installations	Sec. 608
— conductors to be controlled by	531-02
— direct contact protection by, not recognised as sole method	412-06
— earth fault loop impedance for	413-02
— fundamental requirement for	130-04
— HV electrode water heaters and boilers	554-03
— magnetic circuit of	531-02
— magnetic fields of other equipment	531-02
— operating current *(See Residual operating current)*	
— PEN conductor circuits, prohibited in	413-02, 471-08, 546-02
— phase conductors to be controlled by	531-02
— preferred, for shock protection in TT systems	413-02
— preferred, where shock protection by overcurrent devices impracticable	471-08
— prohibited in circuits incorporating PEN conductor	413-02, 471-08, 546-02
— protection against direct contact by, not recognised as sole method	412-06
— protection against indirect contact, in series	531-02
— reduced low voltage circuits with	471-15
— required, for caravan sites	Sec. 608
— required, for socket-outlet circuits in household TT systems	471-08
— required, for socket-outlet circuit supplying equipment outside main equipotential zone	471-08
— selection and erection of	531-02, 531-03
— shock protection by —	
— earth fault loop impedance	413-02
— in IT systems	413-02
— in TN systems	413-02
— in TT systems	413-02
— in TT systems, preferred	413-02
— short-circuit capacity	531-02
— supplementary protection against direct contact by	412-06
— testing	713-12, Appx. 6
Residual operating current—	
— definition	p. 14
— earth fault loop impedance in relation to	413-02
— equipment outside main equipotential zone	471-08
— IT system overcurrent protection, for	473-03
— reduced LV circuit device, for	471-15
— selection of	531-02
— socket-outlet circuits in household TT systems	471-08
Resistance area (of an earth electrode), definition	p. 14
Resistance of insulating floors and walls	413-04
Restrictive conductive locations	Sec. 606
Ring final circuit—	
— circuit protective conductor of	543-02
— definition	p. 14
— test of continuity	713-02
Road-warming cables and conductors	554-06
Rotating machines— *(See also Motors)*	
— emergency switching for	476-03
— exciter circuits of	473-01
— selection and erection of	Sec. 552

S

Safety extra-low voltage, protection by *(See SELV circuits)*	
Safety, fundamental requirements for	Chap. 13
Safety, object of the Regulations	Chap. 12
Safety services,	
— circuits, independence of	Sec. 563
— classification of sources for	Sec. 562
— emergency switching	476-03
— equipment, independence of	Sec. 561
— locations exposed to fire risk	Sec. 563
— omission of overload protection	Sec. 563
— required to operate in fire conditions	Sec. 561
— selection and erection	
— alarm devices	Sec. 563
— utilisation equipment	Sec. 564
— sources	
accessible to skilled persons only	Sec. 562
— equipment supplied by self-contained batteries	Sec. 562
— not capable of operation in parallel	Sec. 565
— operation in parallel	Sec. 566
— single, use prohibited for other purposes	Sec. 562
— with time delay	
— supplies—	
— as fixed equipment	Sec. 562
— assessment	

— batteries	
— independence of	Sec. 562
— preferred protective measures for	Sec. 561
— segregation of circuits	Sec. 563
Safety signs *(See Warning notices and warning signs)*	
Safety sources for SELV	411-02
Sauna heaters	Sec. 603
Schedule *(See Diagrams)*	
Scope of the Regulations	Chap. 11
Screws, fixing, for non-metallic accessories	471-13
Screws, insulating, not to be relied upon as Class II	413-03
Segregation—	
— assemblies of equipment	Sec. 515
— circuits	528-02
— non-electrical services	528-02
Selection and erection of equipment	Part 5
— overcurrent devices for safety services	Sec. 563
SELV circuits—	
— agricultural installations	Sec. 605
— application as protective measure	471-02
— arrangement of	411-02
— bathrooms	Sec. 601
— cable couplers for	553-02
— exposed conductive parts of	411-02
— live parts of	411-02
— mobile safety sources for	411-02
— nominal voltage	411-02, 471-02
— plugs and socket-outlets for	411-02, 553-01
— safety sources for	411-02
— separation from other circuits	411-02
Semi-enclosed fuses *(See Fuses)*	
Separation, electrical, as protection against shock *(See Electrical separation)*	
Services, Safety *(See Safety Services)*	
Shaver supply units in bathrooms	601-09
Shavers, plugs and socket-outlets for	553-01
Shields, fire-resistant, for fixed equipment	422-01
Shields, lampholder, in bathrooms	601-11
Ships, electrical equipment on board, the Regulations not applicable to	110-02
Shock current, definition	p. 14
Shock, electric, definition	p. 10
Shock, electric, protection against *(See Direct contact, Indirect contact)*	
— Short-circuit current, definition	p. 14
Short-circuit current, prospective—	
— assessment at origin of installation	313-01
— at points of protective devices	432-02
— determination of	434-02
— for residual current devices	531-01
Short-circuit, protection against—	Sec. 434, 473-02
— devices for	432-01, 432-04, 434-03
— omission of devices for	473-02
— position of devices for	473-02
Showers, rooms containing, protective measures	Sec. 601
Signal of first fault in IT system	413-02
Signs, electric *(See Discharge lighting)*	
Signs, warning, for areas accessible only to skilled or instructed persons	471-13
Signs warning *(See also Warning notices)*	
Simultaneously accessible parts—	
— bathrooms	601-04
— definition	p. 14
— earth-free local bonding locations	413-05
— SELV circuits	471-02
— with automatic disconnection	413-02
Site-built assemblies, insulation of—	
— as protection against direct contact	412-02, 713-05
— as protection against indirect contact	413-03, 471–09
— testing	713-05
Skilled person, definition	p. 14
Skilled persons—	
— and safety services sources	562-07
— areas reserved for, to be indicated by signs	471-13
emergency switching of safety services by	476-03
— exemption from requirements for shock protection, in locked areas for	471-13
— obstacles permissible for protection of	471-06, 471-13
— placing out of reach permissible for protection of	471-07, 471-13
Socket-outlet, definition	p. 14
Socket-outlets—	
— bathrooms, prohibited in	601-10
— bonding to boxes etc.	543-02
— caravan site installations	Sec. 608
— caravans, in	608-08
— circuits for—	
— disconnecting times	413-02
— earth fault loop impedances for	413-02
— household TT systems	471-13
— supplying outdoor equipment	471-12
— clocks, for	553-01
— construction sites	604-12
— emergency switching by, prohibited	537-04
— FELV circuits	411-03
— functional switching by	537-05
— height of	553-01
— household installations, selection	553-01
— non-conducting location, in	413-04
— outdoor equipment, for	471-08, 471-16
— prohibited—	
— bathrooms	601-10
— 'Class II' circuits or installations	471-09

— emergency switching	537-04
— 'instantaneous' water heaters	554-05
— provision of	553-01
— reduced low voltage circuits	471-15
— safety separated circuits	413-15
— selection and erection of	553-01
— SELV circuits	411-8, 553-01
— shavers, for	553-01
— supplying equipment outside main equipotential zone	471-08
Soil warming, cables and conductors for	554-06
Soldered joints, corrosive fluxes	522-05
Soldered joints, short-circuit with stand	434-03
Space factor, in conduit and trunking	522-08
Special installations needing qualified advice	120-01
Special situations—	
— protection by non-conducting location	471-10
— protection by earth-free local equipotential bonding	471-11
— protection by electrical separation	471-12
Specification, the Regulations not intended as	120-01
Specifications *(See Standards)*	
Spur, definition	p. 14
Standards—	
— British, referenced in the Regulations, list	Appx. 1
Standby supplies	
Starters, discharge lighting *(See Discharge lighting)*	
Starters, motor *(See Motors)*	
Starting currents, assessment for compatibility	331-01
Stationary equipment, definition	p. 14
Statutory regulations *(See Regulations, statutory)*	
Steel—	
— reinforced concrete poles, exemption from protection against indirect contact	471-13
— reinforcement of concrete, use as earth electrodes	542-02
— structural metalwork, bonding of	413-02
Storage batteries for safety services	351-01
Structural metalwork, bonding of	413-02
Sunlight, cables exposed to	522-11
Supplementary bonding conductors, selection and erection of	547-03
Supplementary equipotential bonding *(See Equipotential bonding)*	
Supplementary insulation, applied during erection	413-03, 471-09, 713-07
Supplementary insulation, definition	p. 14
Supplementary protection by residual current devices	412-06, 471-16

Supplies—	
— electrically separated	Sec. 528
— highway power	Sec. 611
— portable equipment outdoors	471-16
— reduced low voltage circuits	471-15
— safety services, assessment	313-10
— SELV circuits	411-02
— standby, assessment	313-02
Supply, nature of, assessment	313-01
Supply systems, public—	
— effect of starting currents on	Sec. 331
— switching arrangements for safety and standby supplies	313-03
— the Regulations not applicable to	110-02
Supply undertakings—	
— consultation with, on starting currents	331-01
— disputes with consumers, procedure	Appx. 2
— equipment, suitability for alterations to installation	130-09
— not compelled to give supply in certain circumstances	Appx. 2
— switchgear, use for isolation and protection of installation	130-06
Supports for cables for fixed wiring	521-03
Supports for wiring systems	521-03, 522-08, Appx. 4
Swimming pools	Sec. 602
Switch, definition	p. 15
Switch, linked—	
— definition	p. 15
— 'instantaneous' water heaters	554-05
— main	476-03
— selection of type	530-01
— step-up transformers	551-01
Switch, main, for installation	476-03
Switchboard, definition	p. 15
Switchboards, conductors on	523-01, 523-03, 514-06
Switchboards, open type, working platforms, etc.	471-13
Switches—	
— bathrooms	Sec. 601
— fireman's	537-04
— linked *(See Switch, linked)*	
— main switch for installation	476-03
— position of, fundamental requirement	130-05
— prohibited in PEN conductors	460-01, 546-02
— prohibited in protective conductors	543-03
— single-pole, prohibited in neutral	130-05, 530-01
— step-up transformers, for	551-01
Switchgear—	
— definition	p. 15
— for isolation and switching	Sec. 537
— for overcurrent protection	Sec. 533
— for safety services	563-01
— for shock protection	Sec. 531
— diagram for	514-09

— linked, selection of type	530-01
— main, isolator for maintenance of	476-02
— marking of	514-01
— overheating and arcing	422-01
— selection and erection of	Chap. 53
— single-pole, prohibited in neutral	130-05, 530-01
— use of supply undertaking's	476-01
Switching—	Chap. 46, Sec. 476
— appliances and luminaires	476-03
— cooking appliances	476-03
— devices, selection and erection of	Sec. 537
— electrode water heaters and boilers	554-03
— emergency	Sec. 463, 476-03
— fundamental requirements for	130-06
— 'instantaneous' water heaters	554-03
— main switch for installation	476-03
— mechanical maintenance	462-01
— of circuits	476-03
— prohibited, in PEN conductor	460-01, 546-02
— prohibited, in protective conductors	543-03
— remote, for mechanical maintenance	537-04
— single-pole, prohibited in neutral	130-05, 530-01
— step-up transformers	551-01
Symbols	514-09
System, definition	p. 15

T

Tanks and taps, bonding of	413-02
Telecommunications circuits—	
— applicability of the Regulations to	110-02
— segregation from LV circuits	528-01
Temperature—	
— ambient, cables and conductors	522-01
— ambient, definition	p.7
— limiting device for water heaters	554-04
— limits—	
— current-carrying capacity	Sec. 523
— heating cables	554-06
— non-metallic boxes suspending luminaires	527-01
— protective conductors	543-01
— short-circuit protection	434-03
— wiring system enclosures	527-01, 422-01
— rating of lampholders	553-03
— surface, of equipment	422-01, 423-01
Terminations of cables, conductors, etc.	Sec. 526
Testing—	Part 7
— alterations to installations	Chap. 72
— assessment of need for periodic	Chap. 73, Chap. 34
— barriers provided during erection	713-07
— 'Class II' enclosures	413-03
— continuity of protective conductors	713-02
— continuity of ring circuit conductors	713-02
— earth electrode resistance	713-11
— earth fault loop impedance	713-10
— electrical separation of circuits	713-06
— electronic devices	713-04
— enclosures provided during erection	713-07
— equipment, fundamental requirement	130-02
— initial	Secs. 711, 713
installation, fundamental requirement	130-02
— insulation—	
— applied during erection	412-02
— non-conducting floors and walls	713-08
— resistance	713-04
— site-built assemblies	713-05
— periodic	Chap. 73, Chap. 34
— assessment of need for	130-10, Sec. 732
— notice on	514-12
— polarity	713-09
— protection against direct contact	713-08
— protective conductors, continuity	713-02
— repeat	713-01
— residual current devices	713-12
— ring circuit continuity	713-03
— test certificates	Sec. 741
— test voltages	713-04 Table 71A
Thermal effects, protection against	Chap. 42
Thermal insulation, cables in	523-04
Thermal stresses of short-circuit	434-01
Thermal stresses on earthing arrangements	Sec. 542
Thyristor drives, assessment for compatibility	331-01
Time delay	
— for safety services supplies	352-01
— for undervoltage protective devices	451-01
Times, disconnecting—	
— equipment in bathrooms	Sec. 601
— of protective devices	Appx. 3
— reduced low voltage circuits	471-15
— shock protection, for fixed equipment outside main equipotential zone	471-08
— shock protection, general	413-02
— short-circuit protection	434-03
TN system—	
— connection of exposed-conductive-parts in	413-02
— definition	p. 15
— devices for shock protection	413-02
— switching of neutral	Sec. 530
TN-C system—	
— connection of installation to Earth	Sec. 542
— definition	p. 15
— explanatory notes on	Fig. 3
— isolation and switching in	460-01
TN-C-S system—	
— connection of installation to Earth	Sec. 542

– definition	p. 15
– explanatory notes on	Fig. 5
– on caravan site, earthing of	Sec. 608
TN-S system–	
– connection of installation to Earth	Sec. 542
– definition	p. 15
– earthing on caravan sites	Sec. 608
– explanatory notes on	Fig. 4
– isolation of neutral	460-01
Tools, for–	
– adjustment of overcurrent settings	533-01
– compression joints	511-01, 526-04
– disconnection of main earthing terminal	542-04
– disconnection of protective conductors	543-03
– opening areas accessible only to skilled persons	471-13
– opening enclosures	412-03, 413-03
Total insulation of equipment	413-03, 471-09
Track system, luminaire *(See definition of Socket-outlet)*	p. 14
Traction equipment, the Regulations not applicable to	110-02
Transformers–	
– current, omission of overload protection	473-01
– of residual current devices	531-02
– safety isolating, for SELV circuits	411-02
– selection and erection of	Sec. 551
Transient overvoltages, assessment for compatibility	331-01
Trunking (for cables), definition	p. 8
Trunking systems–	
– cable capacities of	521-07, 522-08
– exposed to water or moisture	522-03
– fire barriers in	527-02
– joints in	543-03
– partitions in	528-01
– protective conductors formed by	Sec. 543
– selection of	521-05
– socket-outlets in	543-02
– space factors in	522-08
– supports for	522-08
– terminations	526-03
– vertical runs, temperatures, in	Sec. 422
TT system–	
– connection of exposed conductive parts in	413-02
– connection of installation to Earth	Sec. 542
– definition	p. 15
– devices for shock protection	413-02
– earthing resistance requirements	Sec. 542
– earthing resistance variations	531-02
– explanatory notes on	Fig. 6
– socket-outlet circuits in household or similar installations	471-08
– switching of neutral	Sec. 530
Type-tested equipment, Class II	413-03

U

Underground cables	522-06
Undervoltage, protection against	Chap. 45
Untrained persons, the Regulations not intended for	120-01

V

Values of k	Tables 43A, 54B, 54C, 54D, 54E, 54F, 54G
Varnishes, as insulation	412-02, 413-03
Vehicles, electrical equipment of, the Regulations not applicable to	110-02
Ventilation, fixed equipment	422-01
Ventilation of luminaires in caravans	608-08
Vermin, cables exposed to	522-10
Voltage–	
– drop	Sec. 525
– limit	Sec. 525
– tabulated values, for cables	Appx. 4
– exceeding low voltage, applicability of the Regulations to	110-03
– exceeding 250V, warning notices	514-10
– extra-low, definition	p. 15
– for testing	713-04, Table 71A
– high *(See High voltage)*	
– limit–	
– ceiling roses	553-04
– filament lampholders	553-03
– safety separated circuits	413-06
– low–	
– definition	p. 15
– reduced	471-15
– marking, of different voltages within equipment	514-10
– nominal–	
– assessment of	313-01
– definition	p. 15
– reduced low voltage circuits	471-15
– SELV circuits	411-02, 471-02, Sec. 605
– suitability of equipment for	512-01
– oscillation, in IT systems	413-02
– ranges covered by the Regulations	110-03
– rating of fuses and circuit-breakers	531, 532
– tolerances *(See definition of Voltage, p.26)*	p. 15

W

Walls—
- cables concealed in — 522-06
- fire-resistant, cables passing through — Sec. 527
- insulating, resistance of — 413-04
- socket-outlets on, mounting height — 553-01
- thermally insulating, cables in — 523-04

Warning notices—
- caravans, in — 608-07
- earth-free equipotential bonding — 471-11, 514-13
- earthing and bonding connections — 514-13
- emergency switching — 537-04
- isolation of equipment — 514-11
- voltages exceeding 250V — 514-10

Water—
- heaters—
  - and boilers, electrode type — 554-03
  - having immersed heating elements — 554-04, 554-05
  - in bathrooms — 601-12
- pipes—
  - aluminium conductors with — 547-01
  - bonding to, at points of contact — 528-02
  - main bonding of — 528-02, 413-02
  - public, prohibited as earth electrodes — 542-02
  - segregation from — 528-02
- wiring systems exposed to — 522-03

Weather, equipment exposed to — 512-06
Weather, wiring systems exposed to — Sec. 522
Weight *(See Mass)*
Wiring materials, selection and erection of — Chap. 52
Wiring Regulations Committee, constitution — preface p. vii

Wiring systems—
- ambient temperature for — 522-01
- corrosive or polluting substances, exposed to — 522-05
- damage by fauna — 522-09
- installation methods — Appx. 4 Table 4A
- livestock, inaccessibility to — 522-10
- metalwork of, as protective conductor — 543-01
- numbers of cables in — 522-08
- selection of types — Sec. 521

Workmanship — 130-01